I0491493

The Care and Feeding of the Lockheed A-12 Blackbird in Captivity For Museums and Display

A Technical Directive on the Restorative and Display Processes for the A-12 and SR-71

By

Jeannette Remak

Phoenix Aviation Research

We would also like to credit Lockheed Martin for the use of their promotion image of the SR-71 on the covers of this book.

The Lockheed A-12 Blackbird in flight. This is Article #121 tail number 60-6024 (Lockheed)

Contents

Introduction

This book is a revision of the 2017 version. It does contain some new material , updates of various museums, new photos and new material in the Appendices.

The reason for this update is simple; it needed to be done. We have corrected many of the issues the first version had, and the updates were sorely needed. In fact, as the A-12 and the SR-71s in museums, age and some so badly many of the former workers in the programs are feeling their legacy is being destroyed, it is our hope this might generate some more interest in conserving these precious aircraft.

This book is dedicated to the Silent Warriors of the OXCART/ SENIOR CROWN/TAGBOARD programs and the the A-12s and SR-71s that stand for this legacy.

A beautiful restoration job on the A-12 at the USS Alabama, restored after severe damage in a hurricane and displayed perfectly.... Indoors ! (author collection)

The History of the A-12 Blackbird

The Lockheed A-12 Blackbird was the child of the Cold War that erupted between the Soviet Union and the United States at the end of World War II. U.S. political and military leaders feared that the Soviet Union was developing a deadly ICBM arsenal, that it might turn loose on the United States. This belief and the desire to learn more about what was going on in the Soviet Union is what drove the United States to develop more methods of spying on the once WWII ally of the United States. Cold War tensions increased which led to other countries in the Soviet block becoming targets for United States reconnaissance. Airborne aerial reconnaissance was the way to go. It was the quickest, t and quietest means of securing the needed information. The Central Intelligence Agency was the running the intelligence operations for the United States, along with the support of the Air Force. The very distinct tensions and egos that developed between the two services would later become an angry legend in this fight to protect the United States from Soviet attack.

The CIA, while now in the game, did not have the all the experience it needed in creating a vehicle to command the skies for reconnaissance of the enemy. The Air Force did not have the monetary means or the foresight to do it in the black. Hence, the two teamed up for what would be the most incredible aircraft and black program ever created thanks to Kelly Johnson of the Lockheed Skunkworks.

President Dwight Eisenhower, in power at the time and worried about having another Pearl Harbor event hit the U.S., decided early on that he did not want "uniforms" in the cockpits of any reconnaissance bird. *"If uniformed personnel of the armed services of the U.S. should overfly Russia, it is an act of war—legally—and I don't want any part of it"* . Eisenhower did not know, at that point, how right he was. He would soon find out when the U-2 went down over Sverdlosk, USSR. The problem of "uniforms" (aka) military pilots, was solved by "sheep dipping" those military pilots into civilian, non-uniform pilots. The USAF and CIA came together with Kelly Johnson to create the Lockheed U-2 Dragon Lady, a bird that while not having tremendous speed, did have the altitude to protect her. While venerable in service, the U-2 was compromised early in its program with a shoot down over Soviet territory. The CIA had already known that the U-2 did not have a long service life over the USSR and that she would soon be compromised. It meant that there had to be a successor, something that could not be compromised.

That aircraft was the Lockheed A-12 Blackbird and the name of her project was OXCART.

OXCART was one of the CIA's darkest black projects. Devised in absolute secrecy, built in deep black and flown in a place that didn't exist, the A-12 was an aircraft built in the 20th century, that lived up to her 21st century design. To this day, she is still the fastest aircraft used in aerial reconnaissance operations in the world, at least that we know that we know of right now. Both the CIA and Lockheed knew that this was going to be an expensive, high risk proposition, something that neither could fund on their own. Only the government would fund a program like this. Richard Bissell, then head of the CIA Special Projects, looked into what was needed. He decided to appoint a panel of experts to explore this. Dr. Edwin Land, of Polaroid Land Camera fame, became the chairman of the panel that would explore this problem.

From 1957 through 1959, the panel met six times. Attended by Clarence Kelly Johnson of Lockheed Skunk Works, Vincent Dobson, president of General Dynamics (Convair) and the assistant secretaries of the Navy and the Air Force all joined in. The study grew into the project "GUSTO".

GUSTO had many evolutions before it finally arrived at the A-12 configurations Coming from the defunct SUNTAN project, the *Archangel* as the GUSTO bird became known, was a Mach 3 aircraft with a range of 4,600 miles and an altitude of 90,000 to 95,000 ft. Kelly Johnson initially had another idea in mind, the G2A. G2A was a tailless subsonic low RCS (radar cross section) aircraft intended to take the place of the U-2. It was later tossed out because testing showed it might be visible to the Soviet radar system. This left the *Archangel* out there to show what Kelly was thinking about.

This proposal was not without competition. The Convair Corporation (General Dynamics) had an idea too. *KINGFISH* was a parasite aircraft that could be launched from a specially configured B-58 Hustler mother ship. The parasite could reach Mach 4 using a ramjet power. For landing, it would use a more conventional turbojet. As the refining process went on, the design proposals for the Archangel went from A-1 to A-11. While the Lockheed concept was proven by A-11, the RCS (radar cross section) on the airframe was too large. The altitude, range and speed were fine, but it was too visible to Soviet TALL INN radar. The final concept A-12 was the one that won the day. The first mention of A-12 appeared in Kelly Johnson's log on April 21, 1958.

Johnson wrote: *" I drew up the first Archangel proposal for a Mach 3 cruise airplane having a 4, 000 mile nautical range at 90,000 to 95,000 ft. "* Three months later on July 23rd, he followed with; *"I presented this airplane, along with GUSTO model G2A to the program office. It was well received. The navy mentioned in a study they had been making on the slower, higher altitude airplane, on which the program office wanted my comments."*

August 28, 1959, Johnson got word that the Lockheed Skunks Works had won the competition. The next day funding was approved for $4.5 million to cover from September 1,1959 to January1,1960. in September 1959, the CIA gave Lockheed permission to start on the antiradar, structural tests and other functional modifications. It was at this point that GUSTO had become OXCART program.

Building the Blackbird was a work of art. They were built by hand with jigs that were created by hand. This is the production floor of the Lockheed Burbank Plant. (Lockheed)

The Technical Breakdown of the A-12

Construction: The A-12 was designed as a supersonic long-range aircraft, characterized by modified delta, thin, low aspect ration wings. Twin rudders, canted at 15 degrees inward, were individually mounted on each engine nacelle. The wing trailing edges contained inboard and outboard elevons (a combination of aileron and elevator). These features reduced radar signature and aided in controlling yaw in case of a single-engine flight. The A-12 would use two Pratt Whitney J58 engines.

The fuselage, 93% titanium alloy was fabricated primarily out of and used three types of titanium alloy, A-110AT (5A12.5Sn), B120-VCA (113V 11Cr 2Al), and C120AV (6Al 4V). The A-110AT consisted of 5% Aluminum, and 2.5% tin. The B120VCA contained 135 Vanadium, 11%chormium and 3% Aluminum. The C120AV contained 6% aluminum, and 4% Vanadium.

The make up of the titanium alloy consisted of chines located along the sides of the fuselage, airframe components, such as the engine air inlet, rudders upper and lower inserts to the nose section, fuselage and nacelle chines, wing leading edges, elevons and tailbone, incorporated composite material.

The inlet spikes in the forward section of the engine nacelle moved forward and backward to regulate the airflow to the engines above Mach 1.4. The A-12 was flown by one pilot, who would be housed in a pressurized cockpit and dressed in a pressurized, air-conditioned flight suit.

Unexpected difficulties arose from the metal fabrication stage. Titanium was equal to stainless steel in strength, but its virtues as an aircraft metal; light weight, strength, corrosion resistance and high temperatures tolerance were accompanied by new manufacturing. This process required a 200,000 psi forging press that the US company did not have and along with an aging process of 70 hours to bring it to full strength, the product was not acceptable.

With careful aging and quality control, the time could be reduced to 40 hours but that serious glitch appeared with both processes. The titanium being manufactured in the United States in those days lacked the required purity. In technical terms, U.S. titanium was hydrogen embrittled. In simple terms, if a piece of U.S. made titanium dropped to the floor, it would shatter to bits.

The purity problem became a major stumbling block in A-12 production. Initially, all of the manufacturing material secured from *Titanium Metal Corporation* had to be rejected on an contaminated quality basis. The entire first batch of raw material ended up being tossed out, along with the exiting "pickling process" that was used to finish the process. A source of purer titanium had to be found and it would be, outside the United States. The outside source was located—in the Soviet Union. Not only was Soviet titanium of higher quality, but the USSR had the only 25,000 lbs forging press needed to form the basic material. In a remarkable stroke of irony, the CIA was able to purchase titanium from the Soviet Union under covert conditions. The Soviet Union remained unaware that it was aiding in the development of an aircraft that someday might over fly and spy on them.

There were other problems with titanium. It was allergic to just about everything that touched it. Cadmium, mercury, mercury amalgam, cadmium-plated tools, halogens (chlorine, fluorine, bromine, iodine. even ink found in some pens and lead from pencils. Ink from felt tip pens could actually eat a hole in a sheet of titanium in just under 12 hours. Skunk Works engineers found after much detective work, that fabrications and welds, and spot welds done in the summer were more prone to deteriorate than those done during the winter. They discovered that the deterioration was related to problems with algae in the Burbank California water supply, where Lockheed Skunk Works lived. To prevent it, municipal water was heavily chlorinated during the summer. This water was used to wash the titanium plates. The water would literally eat away the welds destroying the plate.

A-12 chine patterns from the Lockheed plant in Burbank (Lockheed)

The airframes could be assembled by conventional construction techniques, but it would take hand–jigging or one by one assembly, to keep the airframe construction process moving. Despite the costs and fabrication problems there was a distinct advantage in using the titanium in the A-12: the hotter it gets, the more the metal "re-cures" itself. That means as heat builds up when the aircraft flies at Mach speed, the metal makes itself stronger, much the way it does in the annealing process.

Notes from the first thermal test state that the wing section *"crumpled up like an old dish rag"* when exposed to the high temperatures of Mach 3 flight. The problem was solved by Kelly Johnson's idea to use corrugation in the test wing section, which would control the shape and direction of the crumpling. When the titanium was heated, the corrugations merely deepened and returned to their original shape when it cooled. This controlled the warping and resulted in the redesign of the A-12's wing to incorporate chord wise (longitudinal) corrugations.

The A-12 had many unique problems with its construction. Kelly Johnson saw unexpected problems with the color coding in the numerous wires and tubes used in the A-12. Lockheed found that ten percent of the technicians working on the aircraft were color blind!

There were separate test units treated to study the thermal effects on the large wing panels. When heated to the temperatures the aircraft could encounter in flight, the panels would warp badly.

It was no wonder there were so many mistakes in the wiring. The team developed odd shaped terminals keyed to ensure that color blind workers could not insert them incorrectly.

Mostly everything for the A-12 had to be re-created even the hydraulic fluid. The A-12 required hydraulic fluid suitable for use at temperatures above 600 degrees F. No such fluid was known to exist. Lockheed contracted Pennsylvania State University for help, and scientists there developed a workable hydraulic fluid by combining some seven chemicals to maintain stability at high, as well as normal temperatures.

These high temperatures also reflected problems in the rubber O rings and gaskets used on the A-12. Fuel tank sealants were manufactured by Lockheed, but when they were exposed to cold temperatures while the aircraft was on the ground, and left for too long, the sealants would revert and turn to something like watery putty. In flight, the fuel tank components expanded from the heat and sealed the system. The engineers never fully corrected the problem of leaky fuels tanks on the ground,: they simply factored it into the fueling and storage decisions. A-12s and SR-71s usually took off with a minimum of fuel due to the leaks on the ground. The leaks were caused by the fuel tanks not sealing properly on the ground and until they had reached Mach 1 flight. The heat generated helped expand the tanks and seal them. Hence the aircraft refueled almost immediately after taking off.

The A-12 did consist of three types of titanium alloys. It also used steel. The corrosion resistant steel used was designated A-286. This was a heat treatable alloy containing about 15% chromium, 26% nickel, 1% molybdenum, and 2% titanium. It could withstand temperatures up to 1,200 degrees. Two nickel alloys—Rene '41 and Hastalloy X—were used in areas that were subjected to intense heat in the engine nacelle ejector section.

The wings on the A-12 are modified, thin, double-delta, fully cantilevered and highly tapered in design. The wing was also designed to be the main fuel cell, holding fuel in the area defined by the leading edge to the elevon support beams to midwing on each side. Both wings support nacelles.

High speed flight was equal to heat build up on the A-12. The wings manage it with corrugated panels that allowed controlled expansion and contraction of the skin. Triangular or V shaped sections made up the leading and trailing edges of the wing. Those wedge shaped panels were filled with composite plastic that absorbed and dispersed radar energy to inhibit to return of radar echo signal. The composite plastic was made from asbestos- silicone laminate and was essentially the same as that used in the chines and control surfaces.

Aircraft #121 and #124 differed from the others in that the fillet panels and skin sections were made of A100AT titanium alloy instead of silicone asbestos. The asbestos components would eventually become an environmental concern during maintenance, as well as during the restoration process later on, long after the A-12s were retired to museums.

The fuel system and the inlets system of the A-12 were extensive and complicated. The fuel had to be delivered and be reliable over intense heat and in a way that did not undermine the aircraft's stability, aerodynamics or small radar cross section.

Pratt and Whitney Engines

The A-12 was powered by two Pratt and Whitney J58 engines, one in each of the two nacelles. The J58 was an axial flow gas turbine with a nine stage single shaft compressor , a can annular combustion chamber and a two stage reaction turbine with afterburners.

A "start cart", powered by two General Motors V-8 racing engines, was employed to get the engines turning for start up. The chemical used in the system to ignite the very high flash point JP-7 fuel was Triethylborane or TEB, which has the unusual property of exploding whenever it comes in contact with air.

The ignition system injected TEB into 16 points in the combustion chamber where the TEB combined with fuel produced a green flash that gave the blackbirds the nickname “green dragon”.

A-12 Cameras

The A-12 camera system was unique. The aircraft carried several different photographic systems.

The Type I camera, built by Perkin Elmer, used a F4.0 18 inch lens and 6.6 inch wide, 5,000ft. supply of film. It could resolve 140 lines per millimeter and provided a ground resolution of 12 inches. The film transport used a concentric supply and take up system to kept the weight of the film centralized; minimizing any shift in the aircraft CG (center of gravity) as the film was advanced. A rotating cube mirror replaced the prism for the scanner.

The Type II camera used a 21” lens and an 8,400 ft, 8” wide film supply. It produced photographic pairs covering 60 mile wide swath with a stereo overlap of about 30% overlap.

The Type III camera was a modified Hycon B camera similar to the one that was used in the U-2. A 26 inch focal length camera was brought on line because the first two systems weren't giving as much resolution as had been planned. The problem with the Hycon was that it really was not made for the high speeds attained by the A-12.

The Type IV camera known as the "big hammer" was an advanced version of the Hycon B camera. It used a 48" f5.6 lens and 12,000 ft of 9.5 wide inch film for extremely high resolution spotting. No long axis camera was ever used in the A-12 as the size created too many installation difficulties.

The A-12 had to be sealed in a special hangar at Area 51 in a virtual "clean room" environment.

The A-12 photo shop hangar had doors that fit tightly around the aircraft so that all dust could be kept our while the film pallets were loaded and unloaded. Before entering the photo shop, technicians had to don complete clean room outfits and go through a high velocity air wash. This procedure was also followed when film was processed at the Okinawa shop where a new facility was built so that film could be swapped out overnight. At the time of the OXCART program, there were actually five Type I camera systems in the inventory. By the time OXCART was phased down, two Type I cameras of the “A” series were placed in storage.

Flight Test The A-12 unofficially began test flights at Area 51 on April 25, 1962 when Lou Schalk(Lockheed test pilot) took # 121 on her first flight—less than 2 miles at 20 to 30 feet altitude. During the short hop, Schalk discovered that the control linkages were not correctly installed. A true maiden flight took place at Area 51 the following day. This flight lasted approximately 40 minutes, during which some chine inserts were lost and had to be replaced.

The first official flight, her true inaugural, also with Lou Schalk as pilot and lasting 59 minutes, came several days later, in the presences of CIA and USAF observers. The A-12 took off at 170 knots, climbed to 30,000 ft and attained a top speed of 340 knots. The A-12 went supersonic during the second official flight, on May 2, 1962.

Four additional aircraft, one of which was a two-seat trainer, arrived at Area 51 in 1962. The J58 engines, however, were still not ready, so early test flights were conducted with J75 engines. As the J58s gradually reached Area 51, the two seat trainer flew with one J75 and one J58. The first A-12 equipped with two J58s was flown on January 15, 1963.

Performance flight test thus began in earnest in 1963, and by July 20, 1963, the A-12 had flown to Mach 3. Before the year was out, the nine A-12s in inventory would make 573 flights for a total of 765 test flights.

On February 3, 1964, the A-12 took its longest sustained flight about Mach 3, traveling at Mach 3.2 at 83,000 ft for 10 minutes.

On November 22, 1963, while the Kennedy assassination unfolded, the A-12 met its speed goals, hitting Mach 3.2 at 78,000 ft. At this time Kelly Johnson wrote in his log:

" The time has come for the bird to leave the nest." It had been three years and seven months since the contract to put the OXCART on the boards had been signed. Kennedy had left us, but the A-12 was truly born.

Modifications -- SKYLARK AND SILVER JAVELIN

Because of the deteriorating political conditions with Cuba, the United States and the CIA had been monitoring the island with periodic U-2 over flights during the summer of 1962.

Those missions revealed development of surface to air missile sites construction. As the summer progressed, military strategists came to believe that the SAM missile site pattern was similar to the layout the Soviets had used at home to defend nuclear missile installations.

On October 14, 1962, U-2 reconnaissance photographs showed clear evidence of long range missiles with nuclear capability on trucks in Cuba. President Kennedy promptly mobilized U.S. forces.

Facing the prospect of a full-scale nuclear war, the Soviets backed down and removed the missiles.

Following the Cuban missile crisis, the United States would continue to monitor activities in Cuba closely. The problem was that the U-2, which had served the United States well for many years, was becoming increasingly vulnerable as surface to air missiles and radar defenses became more advanced and widespread. The A-12 was an obvious and intended replacement, but at the time, everything about it was virtually brand new and untried. This led to Project Skylark, an opportunity to test the plane's operational limits and abilities through possible over flights of Cuba.

Politicians were looking for new ways to keep an eye on Castro without getting airplanes shot down, and the A-12 was looking more and more like the answer. Yet, to upgrade the aircraft and get authorization to fly it over Cuba, the politicians would have to win the support of some key military officials, namely Secretary of Defense Robert McNamara, and much of the USAF brass.

Less than two years after the missile crisis, the National Security Council (NSC) was once again evaluating how it could continue the necessary reconnaissance without sacrificing men and aircraft. In a May 1964 meeting, that included Secretary of State Dean Rusk, and the Secretary of Defense, McNamara, the NSC discussed using the U-2 and electronic counter measures (ECM) over Cuba. McNamara felt that an ECM equipped U-2 would not compromise implementation of SIOP (Single Integrated Operations plan).

Many experts disagreed, contending that the ECM (electronic counter measures) over Cuba would greatly endanger U.S. bombers if they were ever needed for an attack. They also felt that ECM would protect American aircraft flying regularly over Cuba. The ECM protections would be good for the first aircraft over and would not be sufficient support the U-2 on regular missions, as the Cubans would quickly learn to counter the tactic. The estimated chance of a U-2 to evade a shoot down, after the first flight, dropped to 10% assuming the Cubans were determined to get the plane out of its airspace.

What could be done in an all-out effort to get OXCART ready? John McCone, then CIA director, answered that the problems were being worked on as they rose: he promised to keep the pressure on, but didn't and that was acceptable. They decided that the A-12 was the answer to their reconnaissance problem and the CIA and Lockheed were to get the A-12 ready as fast as possible. This was s the beginning of Operation 'SKYLARK'.

In a memo dated August 22, 1964, acting CIA Director Marshall C. Carter told the A-12 development team that the plane was to be ready for a Cuban over flight mission no later that the week of November 5, 1964. The memo also laid down the flight characteristics needed for the mission: Mach 2.8 with an altitude of 80,000ft and a range of 2,500 nautical miles or better. It would include four OXCART aircraft. Carter went on to state that Operation SKYLARK was to have the highest priority, unhampered in any way by contractors, commanders, or any other entity that would have a direct effect on the completion of the program objectives. It was going to be an "all out, no holds barred" effort.

At the time SKYLARK was being worked, the longest sustained A-12 flight with two J-58 engines was 4:25 hours. The trainer, #124 held the longest sustained flight record with two J75 engines: 5:25 hours.

The top speed the trainer reached was Mach 3.27, and the maximum altitude attained was 85,000 ft., the longest sustained flight to date that closely approximated design conditions. Using the wind tunnel tests, A-12 engineers were able to improve the aircraft inlet recovery and distortion to specification requirements to maximize range and engine life. Yet major problems remained with fuel consumption during the climb to altitude; the fuel was insufficient to meet the specified cruise range. Quality assurance was a major concern in the SKYLARK project. There were many issues at hand, false cockpit instrument readings, fuel tank leaks, hydraulic leaks, and pressure fluctuations in the brake system, compressor disc durability problems involving excessive growth after repetitive cycles to Mach 3.2 were showing up in the ground test as early as 1964. Meanwhile, the flight speeds were restricted to Mach 2.8, which limited the maximum temperature and steep thermal gradient imposed on the disc by rapid descents from hot to cold environment.

Because of all the corrective measures taken in 1964, the 18 flight engines plus all the new production engines, acquired new discs and were not restricted in any way. Other problems showed up in SKYLARK, engine nozzle actuator pipes failed twice during flight. This prompted an investigation, which showed vibration and system instability problems. Those were addressed at the Area 51 test site. Replacement pipes were sent to the site and a search began for the cause of the vibration. Because of risks and demands involved, pilot comfort was another concern. Project works developed a new parachute pack, lighter and 1 1/2 inches thinner than the previous one. This pack, with better seating, would allow the pilot more mobility on the long missions being planned.

Much of the inertial navigation system INS and the ARC-50 aircraft communications system were revamped. Detachment pilots helped conduct continuous AR–50 test, ranging from 500 nautical miles down to one nautical mile, and automatic direction finder (ADF) test from 200 nautical miles to a contact point.

The CIA report on SKYLARK included a test mission summary. Since the first flight of the A-12 on April 26, 1962, 1,234 flights had been made, totaling 1745 hours; All the flights were done by the 13 aircraft in residence at Area 51. Of the totals, 794 flights accumulated 104 hours using the J58 engines. The maximum speed encountered was Mach 3.27; maximum altitude was 85,000 ft. Seven aircraft, including #124, (the trainer) were assigned to the detachment and were flown by operational pilots. Four of the seven aircraft were primary Skylark aircraft and included #125, #127, #128 and #132.

Four of the assigned to flight test while two #129 and #131 were assigned to detachment as operational aircraft, after modifications were installed.

By January 27, 1965, #129 had completed the first in a series of long range, high speed flights. This project within SKYLARK to demonstrate the A-12's maximum range was known as SILVER JAVELIN. The total flight time was 1:15 hours above Mach 3.1, with a total range based on final flight data, of 2,580 nautical miles at a cruising altitude between 75,600 ft. and 80.000 ft. This was the longest sustained flight bordering on design conditions. Before #129 took the second SILVER JAVELIN flight in early March 1965, Lockheed made a number of modifications to the aircraft, including additions to the air inlet duct seals to improve inlet efficiency strengthening the rudder actuator linkage, and rescheduling the fuel management system to keep the aircraft balanced and reduce drag.

There wasn't much flying done while the Phase II SKYLARK was being put into place. At no time were there less than five operational aircraft available for service.

SKYLARK Phase II also encompassed work on the J58 engines; it brought about a new inlet system, auto forward bypass, J-cams, and duct seals. Designers also improved the composite panels and use the Blackbird's signature black paint on the complete aircraft for the first time. There was improved nitrogen conservation, rudder improvements to support the 450 KEAS climb, film and map destruction capability and the first incorporation of the "Birdwatcher" System (electronic means of tracking the aircraft on a flight on a separate frequency).

The Phase II summary of the aircraft performance showed that the A-12 had achieved Mach 2.9 with a range of 1700 nautical miles from tanker to tanker hookup, with an altitude of 76,000 ft in test. The projected performance for the A-12, after the Phase II mods were completed, was flight speed of Mach 3.05 with a range of 2,500 nautical miles and an altitude of 76,000 ft.

The minimum modifications that had to be in place upon completion of Phase II need to support SUPERMARKET (the ECM package), incorporated in the Lockheed inlet control design and provided for the three–refueling mission. The actual overall mission reliability for Cuban over flights, there as still concern for the ECM problems and fear that those problems were being resolved too slowly. The A-12 might be able to overfly Cuba, but it would have to do so without ECM protection.

The training missions continued for nearly three months after SKYLARK and demonstrated that the A-12s were capable of collecting photographic reconnaissance. But due to the delicate nature of the negotiations surrounding the crisis, the A-12s were never used over Cuba. Instead the U-2s continued to photographs Castro's installations and the breakdown of those installations just as before.

Completion

There were still some modifications going on to complete the bird. KEMPSTER A-B was an RCS (radar cross section) device that utilized different devices on A-12, which was generated by an electron gun, which produced a cloud that could absorb radar frequencies. The equipment was held in the Q bay. The ion guns created a stream of ionized particles that materialized from small holes in the chines, which was just ahead of the air inlets.

The concept was solid and had the A-12 remained in service and not been cancelled for the SR-71, all A-12s would have carried the system. In a restoration process for the A-12 #122 at the Intrepid Museum , the author found the small holes in the chine at station #715 on the aircraft.

The other program EMERALD ,was the development of device which would generate a seeded plasma electric arc for the purpose of absorption of radiation. EMERALD was not continued.

The CIA began to qualify pilots and ground crews on a series of training flights that simulated missions over Cuba. The training mission continued for nearly three months after SKYLARK and demonstrated that the A-12 was indeed capable of collecting photographic reconnaissance. But due to the delicate nature of the negotiations surrounding the Cuban crisis, the A-12s were never used over Cuba. The A-12 was now awaiting a mission of her own.

In 1965, the United States began sending troops to Southeast Asia and Vietnam, moves that created a greater need to substitute the A-12 in place of the U-2. This was needed because of the vulnerability of the U-2 due to SAM sites newly placed in the Vietnam, Southeast Asia area.

New CIA director, William Raborn, said the A-12 could operate in the new theater once final operational readiness tests were passed. Once again, due to politics prevalent in any project even black ones, the 303 Committee, which was created by the NSC (National Security Office) to oversea covert activities, turned down the A-12 for deployment on the basis that the Japanese might find out about it and object, something at the time that was not wanted.

On August 12, 1966, President Lyndon Johnson upheld the 303's decision. The CIA then asked for over flights on Cuba. Again the 303 withheld its agreement, again because they feared that it might upset an already fragile peace in the area.

On May 16, 1967 President Lyndon Johnson finally agreed to let the A-12 be used to see if there were any SAM sites that had gone undetected in North Vietnam. By May 17, 1967, the airlift to Kadena Air Base, Okinawa, Japan had begun, for what was to be known as Project BLACK SHIELD . The first movement of A-12s to Kadena Air Base transpired May 22, 1967. Thirteen days later, the A-12 OXCART was ready to fly its first mission.

On May 31, 1967, the first of the Black Shield missions was flown. This included one pass over North Vietnam and another over the DMZ. The A-12 flew at Mach 3.1 and 80,000 ft for a flight of 3 hours and 39 minutes. It photographed 70 of 190 suspected sites and 9 other priority targets. The A-12 did not detect any radar signals bouncing off her during the mission, meaning her mission went undetected by the Chinese and the North Vietnamese.

From June 19 through August 21, 1967, seven more Black Shield flights were made, with 14 more missions from August 31 through December 16. On the December 16 flight, there was one latch-on by Chinese Fan Song guidance radar, but it was not successful.

OXCART's efforts and results in Vietnam were truly stunning, since it was the first time that non satellite reconnaissance at high speed and altitude could be maintained without the worry of being shot down. Flying over enemy territory was just about what the A-12 was born and bred to do. There were 22 Black Shield missions and the A-12 flew all of them returning unscathed. The photographs these flights produced provided exactly the kind of information the military and the CIA wanted. The A-12's cameras snapped pictures of airfields, military hardware and military infrastructure.

On October 28, 29 and 30 1967, Black Shield flew missions that covered about 55 percent of North Vietnam, including Hanoi, Hai Phong, Pingshiany, and the Dong Dang area. The flights surveyed all six of the major airfields of North Vietnam and more than half of the SA-2 SAM sites.

The flights were used for bomb damage assessment and to search for surface to air missiles. Missions 6732 and 6734 of October 28 and 30 1967 involved two passes over North Vietnam panhandle. The 6732 mission passed along the Chinese border. Mission 6733 on Oct 29th also flew along the border, and the combined reconnaissance from all three missions yielded no evidence of surface to air missiles.

These three missions photographed more than 260 SA-2 sites in North Vietnam, including two new ones never before seen. The SAMs were plaguing U.S. pilots at every turn and the more that could be discovered in their hiding places, the better.

The A-12 was also giving good assessments of the damage from carpet bombing that B-52s were carrying out daily in North Vietnam. The photographs also revealed the carpet bombing along the lines of MiGs left on these air fields had not done nearly as much as expected. The last of the Black Shield missions was flown on May 8, 1968.

The Close of the OXCART program

As early as September 29, 1966 the Deputy Secretary of Defense Paul Nitze proposed at an executive committee meeting that OXCART be phased out of the inventory. He was ready to discuss the USAF's version of the A-12, the SR-71, a slower, two man aircraft without the speed and camera capabilities of the A-12. The SR-71 was in advanced tests and true to the USAF penchant for wanting everything, they wanted their position back as the ruler of the skies.

The A-12 had taken that away from them via the CIA. This was something that ate at the USAF's ever increasing ego and budget. In Nitze's mind, and many others, two supersonic spy planes was one plane too many.

The United States didn't need it nor could it afford it. Dr. Alexander Flax, director of the National Reconnaissance Office (NRO), had circulated papers showing that the SR-71 was in satisfactory condition and could take over North Vietnam duties as soon as December 1, 1967. The Joint Chiefs of Staff concurred that the SR-71 was ready to go operational. But one member of the panel, Dr Donald Hornig, disagreed based on the calculations on equipment lists, statistical factors, and performance curves. Hornig stated that the SR-71 was two to four times more vulnerable than the A-12. Hornig looked at the SR-71s operational techniques and impact including the ECM systems and abilities, present enemy activity and perceived future operations. Hornig concluded that the committee shouldn't be too quick to deploy the SR-71. Dr. Flax pointed out that is there weren't any economic constraints, he would keep the entire force, but money as always was a factor. Flax also felt that there was a need for a firm decision.

If no decision was made, keeping both programs would cost around $32 million. The question of putting A-12s in storage and later retrieving them was also raised. It would cost $300,000 to $500,000 in 1968 dollars to reactivate each aircraft if they were put away and then brought out for flight status again. This cost based on reactivation being done within the first year. The financial plan in place at the time would fund the A-12 and allow continued operation at Area 51 only through December 1966, any longer would require additional funds.

Some members felt that a delay of three to six months in shutting down OXCART would be appropriate. Even Dr. Flax felt that the delay would give a higher degree of confidence that the SR-71 would be able to carry out the job in the face of North Vietnamese defense improvements. Deployment, Flax felt, should be delayed for three months and the deployment should be scheduled for February 1968.

This meeting set the stage for the closing down of the OXCART program. It would be a matter of when it would be convenient. The SR-71 did not reach operational deployment at the expected time and the A-12 was extended through March 31, 1968. OXCART was extended again through June 30, 1968 allowing a one month overlap for the SR-71 to take the field.

There was a test program run between the SR-71 and the A-12. It was called NICE GIRL. This allowed for both aircraft to follow a complex and distant flight plan. As it turned out, there was no competition between either aircraft, as no one could be declared the winner, it was a draw. The demonstration didn't change the minds of the people in charge. The A-12 was available, but the political will wasn't. The USAF had been supporting OXCART since the inception, but USAF pilots weren't flying it, CIA pilots were. The A-12 was being serviced on the ground—and refueled in the air---by the USAF, but the USAF was not getting the glory of performing the mission. The USAF was not comfortable as the support team for the A-12 in a black project headed by the CIA. In short, USAF wanted the CIA out of the aircraft business.

Due to overlapping capabilities, the purchase of the SR-71 by the USAF effectively bought the A-12 off the flight line. As always, there was another political gain to be had. The purchase of the SR-71 gave the USAF the lead in aerial reconnaissance once again. The A-12 was operated by a civilian agency and could be deployed quickly and with a minimum of logistics and that stuck in the USAF's craw. The A-12 had to be retired.

That is how it ended, one of the most spectacular programs in aviation history. OXCART was brought down by ego, politics, and petty jealousy. The OXCART program had so much life left in it, but it was relegated to a hangar to save the face of the USAF that was already overblown with post war greed.

1993 thru 1995
Damage to Nacelle.
Also note that the spike
is fully extended

Damage to A-12 #122 during the years of 1993-95

(Author Collection)

1993 thru 1995 ,
I caught this individual sanding the composite vertical.It was a surprise to the USAF Museum, too.To this day, the vertical can never be made to look "right"

The photo comes from a 10 year study that I created of the A-12 #122. Here is one of the worst things that you can do to an A-12 vertical. The vertical on #122 is made of composite material NOT metal. It was never possible to repair it properly after this treatment. It still shows damage after a superb paint job (Author Collection)

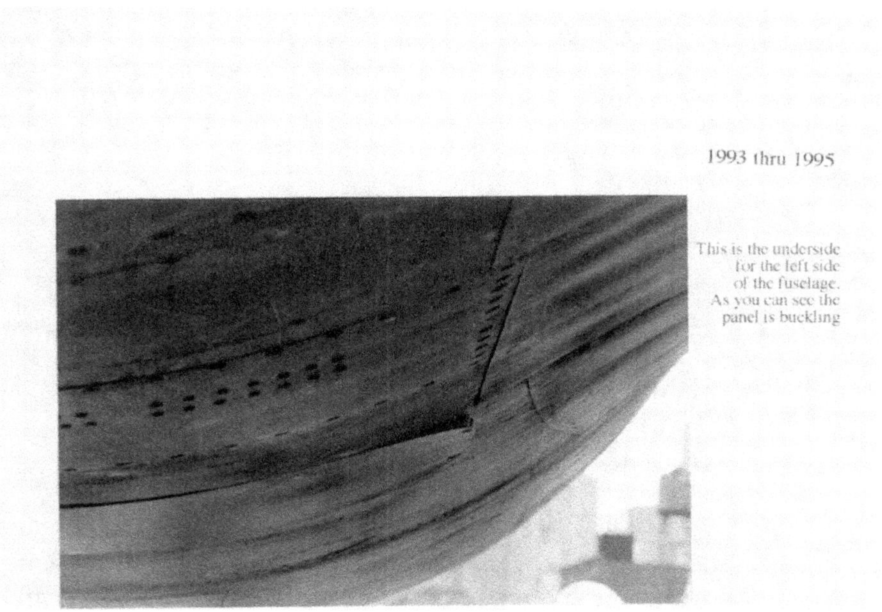
1993 thru 1995
This is the underside
for the left side
of the fuselage.
As you can see the
panel is buckling

Another image of #122 before restoration in 1996, the panel is buckling (Author Collection)

A-12 #122 traveling up from the south in ocean water which is of course, salty. That is one of the problems with transporting an aircraft in this method. The aircraft should have been washed down with distilled water on arrival at the Museum, but that was never done. (Author Collection)

U.S. AIR FORCE

A-12 #927 on display at a museum. This aircraft is the ONLY trainer model for the A-12s. This is a very unique and rare aircraft left to the elements. (Author Collection)

A-12 #938 on display. This exquisite restoration was done by the Museum's curator. The aircraft was a victim of a hurricane . (Author Collection)

AEROSPACE
HISTORICAL
CENTER

Another A-12 on display outside the museum on a pedestal left to the elements. (Author Collection)

Completed restoration, beautifully done by Curator of the USS Alabama Museum. After the aircraft was hit by Hurricane Katrina. The curator hand restored this aircraft manufacturing the parts needed to complete this A-12. (Author Collection)

An exquisite design for display by the CIA of the A-12. This aircraft was actually mishandled at another museum and recalled to become the exhibit for the CIA who flew the A-12 in Operation Black Shield. However, outside display is a bone of contention because many of the museums that do display outdoors, do not care for the aircraft properly.
(CIA)

The same aircraft (USS Alabama Museum) shown prior in this book, took a severe hit by Hurricane Katrina. The Museum Curator did all the work by hand following the A-12's manuals to recreate parts. (Author Collection)

The A-12 and the SR-71 on display in California airpark both looking good due to the arid conditions.(Author Collection)

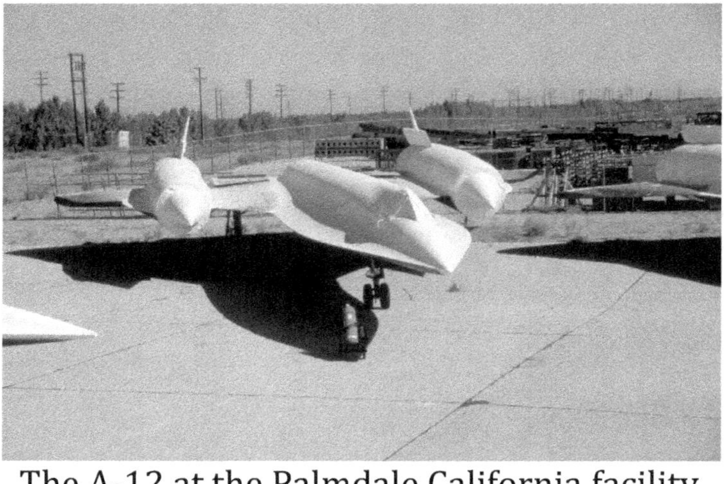

The A-12 at the Palmdale California facility wrapped in Spraylat, also known as latex. (Author Collection)

Asbestos leaking through an A-12 that is displayed outside (Author Collection)

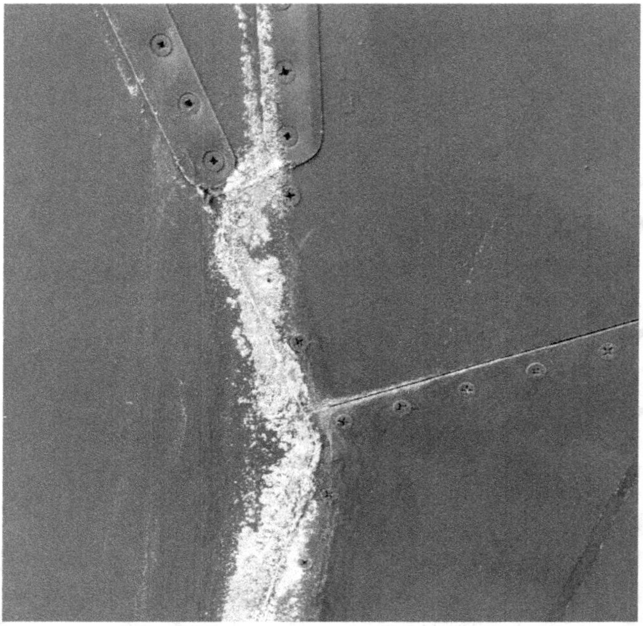

This is an example of *Kadena Krud*, an algae that actually grows on the metal of the A-12 which is 93% titanium. This occurs when an aircraft is outside and subject to rain and the elements. It can be controlled but once it starts, there is no going back. Any source of moisture will cause it to return. This problem also occurred when the aircraft was in service during Black Shield at Kadena Air Base in Okinawa and is prevalent in aircraft that are displayed outdoors (Author Collection)

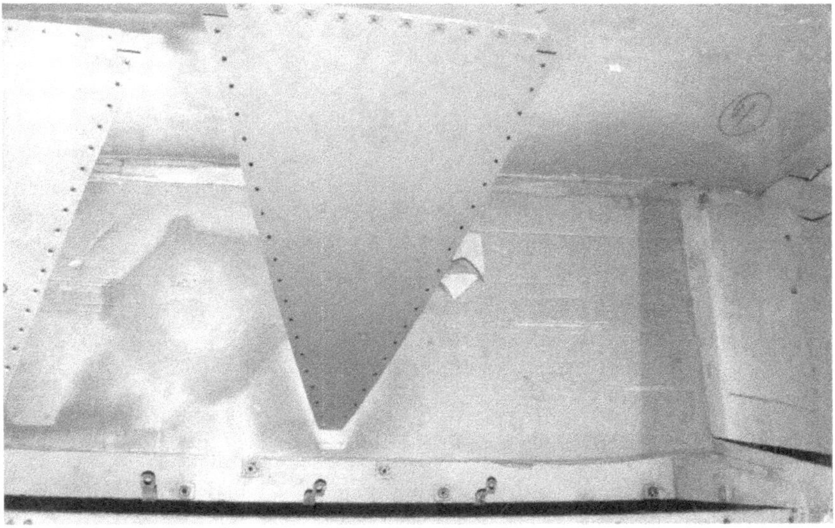

Saw tooth panel from A-12 that had been
removed because of asbestos and recreated in
sheet metal to be
replaced on the aircraft #122 (Author Collection)

Restoration of A-12 #122 tires. These are new and filled with foam so that they will not sag, disc brakes were removed, restored and repainted. Tires must be rotated every couple of years so that dents do not form in the tires.
(Author Collection)

The Bypass panels were removed and sealed so that water would not enter the A-12 #122, this was a critical part of restoration. The aircraft could no longer be subject to water damage internally. (Author Collection)

What is a Museum?

The technical description of a museum is: "*a permanent institution in the service of society and of its development, open to the public, which acquires, conserves, researches, communicates, and exhibits tangible and intangible heritage of humanity and its environment for the purposes of education, study and enjoyment. " (Int'l Council of Museums*). When it comes to Military Aviation Museums, things are just a little different. Basically, not only are they the keepers of history for the aircraft, it is a hard look at the handmaidens of war and where they fit in the world. A Military Aviation Museum is unique in many ways. While an art museum brings you the beauty of a painting or a sculpture, an aviation museum brings the beauty of line, form and design and the power that an aircraft does possess, and also her deadliness.

You see the beauty but you also see the deadly purpose of a machine of war. Military Aviation Museums show the public where their tax money is going and explains the necessity of these awesome aircraft in the defense and protection of the nation.

Responsibility of a Museum

The responsibility of a Military Aviation Museum is also unique. The stories that these aircraft hold are powerful. It is a monumental task to present these aircraft in the correct light with the correct history, and not to be displayed as some oddity or menace, but as the entities they are. According to the American Association of Museums:

"Museums in the United States are grounded in the tradition of public service. They are organized as public trusts, holding their collections and information as a benefit for those they were established to serve. Members of their governing authority, employees, and volunteers are committed to the interests of these beneficiaries. The law provides the basic framework for museum operations. As nonprofit institutions, museums comply with applicable local, state, and federal laws and international conventions, as well as with the specific legal standards governing trust responsibilities.

This CODE OF ETHICS FOR MUSEUMS takes that compliance as given. But legal standards are a minimum. Museums and those responsible for them must do more than avoid legal liability; they must take affirmative steps to maintain their integrity so as to warrant public confidence. They must act not only legally but also ethically."

That is the magic word, ETHICS. In its responsibility to the public and the Museum itself, ethics is the principal function. Without it, a museum is no more than a store house or a side show.

The A-12 in the Aviation Museum system

In the early 1990's, the A-12 Blackbird was tossed out into the public eye without any fanfare or any concept of just who and what these magnificent aircraft really were.

As the Nat'l Museum of the USAF Program Coordination office took possession of the seven Blackbirds, no one truly understood their rarity, history or accomplishments. It should be noted much of that happened because; the program was so black no one knew outside of the CIA who had not declassified the program.

There were 7 aircraft that looked like the SR-71 but weren't and no one had handle on the fact that they came without any manuals, descriptions or nomenclature. As host museums lined up at Program Coordination doors to adopt one of these fine aircraft, the usual credentials were laid out to see if the host museums would be able to meet the criteria. One by one they applied and the bidding war began.

At first, the New England Air Museum was the first to put its bid in on an A-12 #122. $250,000 in escrow had been held to show their intentions were good and that they could comply with the loan agreements of the Nat'l Museum of the USAF contract. It would be a handsome aircraft for the east coast museum, who could prove not only were they competent enough to handle it , they had an indoor spot that would be perfect for the titanium skinned A-12.

Other museums followed, Alabama, Minnesota, San Diego, they all came. The deals were done and the NEAM on the east coast, would get their A-12 until something queered the deal . Queered it was and that honor belonged to no other than the late Zak Fisher, owner and operator of the Intrepid Museum in NYC. While that was good for Intrepid, Zak Fisher's wheeling and dealing within the military circles was well known. A real estate king in NYC, he had the money and the contacts to make his dream of an SR-71 come true. Fisher wanted a SR-71. He had no idea what an A-12 was, except that it looked like an SR-71. Fisher, with a little help from the Pentagon, managed to pull the NEAM's A-12 right out from under their noses and hauled over to Intrepid. It needs to be noted that Fisher and the Pentagon jumped all over the Program Coordination office at Wright Patterson AFB and the Nat'l Museum of the USAF director to get this deal done. The ill equipped Intrepid operations department, with not an aviation professional among them, transported the unprepared A-12 through the open sea on a barge back to the Hudson River and home to Intrepid.[2]

How not to transport a titanium alloy aircraft. This is how A-12 #122 was brought up from the west coast into the Hudson River in NYC. (Author collection)

[2] For the entire story of the A-12 #122 saga see: "The Blackbird's Tale-How an Aircraft Suckered me into things I never believed possible with the help of an Aircraft Carrier" – Jeannette Remak Phoenix Aviation Research 2017

Why the Loan Agreement doesn't work

The Nat'l Museum of the USAF loan agreement that supports the transfer of aircraft from the Nat'l Museum of the USAF system to a host museum is adequate itself as a written document. The problem with the loan agreement is adherence to the policies set in it. The Program Coordination office, who oversees the loans cannot in reasonable thought manage to look after the hundreds of aircraft that they have on loan to host museums. Hence, the policies they put forth can't be supported and the breakdown of discipline regarding the care of the aircraft happens. The loan agreement requires a yearly update on the aircraft including photos. This means that over a year a complete check of the aircraft must be made.

Any maintenance issues, repairs, or new damage must be reported to the Nat'l Museum of the USAF. There is no way that this will happen. Only the most professional of museums can uphold this end of their loan agreement.

Breakdown of a Museum Procurement and loan system

What needs to happen for the many aircraft in these loan programs to be successful is for the Nat'l Museum of the USAF officials to visit the aircraft at least every 2 years. However there is not enough money or manpower for that to happen. Since this can't happen, there is no way to police the various loans out there. Without this type of support, there is no way to check on the various loans and the host museums are left to their own devices.

An attempt has been made by the services, at least in military museums to sort of police their own: for example, the Navy Museum would notify the Army if they saw something wrong with an Army exhibit etc.

It may solve some of the problem, but in most cases it does not. The loaning of aircraft via political pressure needs to also be controlled. This is another case of jumping the gate to get what you want. Using politics and pressure in attaining aircraft for no other reason than to satisfy a want political or otherwise, is dangerous in the Museum loan policy program.

Once this sort of tactic is allowed to happen, it is hard to stop, because a precedent has been set. In the case of Intrepid, this aircraft has undergone the worst of maintenance, in the worst possible place for an aircraft of its type, the flight deck outside seventeen stories about a brine river, the Hudson, in NYC. And, because of political pressure, the Nat'l Museum of the USAF was not able to effectively protect the aircraft it loaned out. The ethics of Museum loan policy was brushed aside to the detriment of the artifact.

Non Vetting of a Host Museum

One of the requirements of the loan agreement is to show that there is a maintenance plan in place, a safe exhibition space and money enough to support the aircraft once it arrives at the host museum. Adherence to that requirement is also a common joke among many of the host museums. Because there is not a proper policing policy and no way of disciplining the host museum for non-compliance, many of the poorer quality museums sneak past in their duties regarding care for their artifacts. Host Museums, in general, need to be gone over with a fine tooth comb to prevent the type of mishandling that has caused the destruction of some aircraft by neglect. .

There must be a plan in place that can be proved viable before the loan agreement is signed. As has happened in the past, because of either political concerns or past history with the lender museum, other aircraft are usually given over without much thought to its care.

There has been some minor cause for joy in this issue. The *Memphis Belle*, a famous B-17 that had been in a museum which couldn't care for the historic aircraft, had been recalled back to the Nat'l Museum of the USAF for restoration and exhibition.

The restoration has since been completed and is magnificent! However, the only way this occurred was because the Nat'l Museum of the USAF had more clout politically than the host museum. The National Museum of the USAF now has the aircraft on display. This doesn't happen all the time and when it needs to, and that is the major problem: There is often NO WAY to recall an aircraft that is being mishandled back to the Lender Museum. Which is the reason why so many aircraft are left at the hands of museums that just don't have the funds or wherewithal to care for precious exhibits.

How to care for an A-12 ---- Fact from Fiction

The A-12 Blackbird, as it's been show in the above paragraphs is not your usual aluminum skinned aircraft. This aircraft, by its nature, should be an indoor exhibit which will protect it from harsh outside environment. That basically means that life on a flight deck, 17 stories above a brine river and inside a known hurricane zone is not the best of all showrooms.

While the A-12 is a tough "bird" in many circumstances, it still needs maintenance, painting and repairs all through the year if you wish to keep it in the prime of condition. That does mean setting up a rigorous inspection and repair plan along with keeping an exhibit area that can best show off the aircraft and protect it from further wear and tear.

The better kept, the fewer headaches. It also means that there needs to be enough money is the restoration budget to care for such an exotic aircraft.

HOMES AND HORRORS

Neglect by some host museums and care by others.

If you take a look at the various museums that house A-12s and we will include SR-71s, some of them don't worry too much about their charges or their special care.

CASE:

1. *Seattle Museum of Flight—Washington*: This aircraft is kept indoors in a magnificent display and kept with exemplary care. The A-12 configuration with the D-21 drone mounted on top is rare. Years of work, with volunteer efforts and historical research, went into preparing this aircraft. This exhibit shows what can happen when the right museum knows how to handle precious artifact.

2. *The San Diego Air and Space Museum:* This aircraft has been left in an outside display. Actually, the aircraft is treated more like a gate guard at the entrance of the Museum. No care has been given to this aircraft for some time and it is quite evident that the lack of care is showing. The aircraft is weathered and covered in bird poop, which because of it acidity can cause lots of damage. Since the Nat'l Museum of the USAF has not been out to visit this aircraft or even consult with the aircraft curator concerning it, this is a prime example of how a loan agreement means nothing. The aircraft, subject to hostile weather and environmental conditions will continue to deteriorate until someone informs the Nat'l Museum of the USAF of

its condition. That report, can only be done by concerned museum visitors, who visit the museum in question, and notate the problem. The museum visitor's effort in reporting it either to the host museum itself, or to the Nat'l Museum of the USAF could help. Yet, there have been reports of other visitors who have taken the museum to task concerning the A-12, but they have all but been ignored. That is a cause for concern not only for the A-12s, but for the SR-71s of which many are constantly being complained about due to lack of care. In the case of the Seattle Museum of Flight, this is how an exotic artifact should be cared for and is an example to every other museum on what to do. In the case of the San Diego Museum, the Nat'l Museum of the USAF should have been alarmed at the aircraft's state when notified by patrons. However, since the Nat'l Museum of the USAF has not been out to inspect the aircraft, the museum continues to proceed as it always has and ignore the situation. The case of the Intrepid Museum is on record here. It is the same political environment and whether things have been successfully worked out remains to be seen. The

Alabama Space and Rocket Museum in Huntsville, was marked up as a NASA aircraft and was shown as a SR-71. To this date nothing has been done by the Nat'l Museum of the USAF to correct the situation even after the attempt of a recent restoration, which did not include buttoning up the aircraft against weather elements and the incorrect blue paint job.

3. *The Battleship Alabama Memorial Museum:* This A-12 has always been indoors. During Hurricane Katrina, the aircraft was badly damaged. The aircraft curator restored this aircraft by hand with every detail lovingly placed. This aircraft was the showpiece for the "Welcome Home Roadrunners -2008" celebration. This again shows that even with little or no money, miracles can happen if you have a museum that cares.

4. *Alabama Space and Rocket Center:* This is again an outdoor exhibit. The aircraft has been recently been "restored". The aircraft is now sporting a blue tone to its paint scene because someone at the museum decided to follow a misinterpreted quote from a Kelly Johnson biography. This has been a major problem in

the Blackbird world of fans and admirers. Not only has the aircraft been previously marked as it flew for NASA, which it never did; it is also displayed as an SR-71 which is completely erroneous. This museum is also the home of the Space Shuttle camp for kids. This is another case of the Museum system turning a blind eye as to what happens to aircraft after the loan agreements are signed. There have been calls by many concerned aviation buffs and former program workers to the Nat'l Museum of the USAF regarding the problem of how the restoration was carried out just recently. Nothing was resolved.

5. *CIA-Langley VA;* This is an outdoor exhibit on a pylon of an A-12. This aircraft was recalled from the Minnesota Air National Guard Museum and given to the CIA. The reason for this was political, but also solved the problem of a rather dubious individual who claimed the aircraft for his personal use. Again, this was a long term problem that was known by the Nat'l Museum of the USAF but nothing was done until it was

almost too late. Due to the CIA request for an A-12, this did solve a very touchy problem.

6. *Blackbird Park- Palmdale California:* This is another case of an outside aircraft. However, in this case the weather is not a problem as it is warm and dry. The aircraft does see routine maintenance and is cared for. However, there is very little if any communication between the Nat'l Museum of the USAF and the airpark

7. *Intrepid Museum-New York City*: the aircraft is outdoors and subject to repeated negligence. Weather varies from ice, freezing temperatures, wind and snow in the winter to humidity, heat and acid rain in the spring fall and summer. It is only because of harassment of the Nat'l Museum of the USAF by various sources, that some of the negligence and vandalism has been addressed. However, since the refit of the Intrepid herself many years ago, the aircraft is still questionable as to the depth of resolve in care and maintenance, since the museum recently again acquired a space shuttle and it has turned its financial attentions to

building a tent to display this item. Once again, damage was done to the shuttle and the tent when Hurricane Sandy came to NYC.

Out of seven aircraft, the only remaining A-12s in existence, there are:
2 in poor condition
2 in political situations dictating their care
3 in good museums that are being cared for

The SR-71 list on display is large. It will be listed in the **Appendices** along with a list of the A-12s. Regarding the condition of these aircraft, the majority of them with the exception of Duxford, UK, the Nat'l Museum of the USAF, Blackbird Park, California (outside), the Smithsonian Air and Space Museum and a couple of others, are inside display only. These aircraft are kept in lovely condition.

Case in point: #122 and the Intrepid—a history of trouble and years of neglect

The A-12 on the flight deck of the Intrepid Museum in NYC had been on that outside deck since 1992. That is a lot of acid rain, heavy snow, ice storms and rain, not to mention Nor'easters and hurricanes. The aircraft on Intrepid has seen only 3 serious attempts in almost 15 years at serious restoration work in all of that time. The first was to seal the many open areas and repair damage to the chine areas along with removal of asbestos from the saw tooth panels on the wings.

There was a monumental amount of work done with care and has kept the aircraft safe from further interior harm all these years. The tires were changed for new ones and filled with foam so they would no longer go flat. The next attempt and that is all it was, was a paint job that was done over the existing paint job causing a pillow effect because moisture had gotten into the new paint coat.

The paint did not adhere properly, hence the pillow effect. The aircraft was also subject to more damage by mishandling. The aircraft sustained holes in her side due to being slammed up against a wire fence on the deck of Intrepid. We might also note that while moving an aircraft around is at best a dangerous assignment, it is even more dangerous in the case of the A-12 due the her unique configurations. You don't ever attempt to back it up, as some have done.

There was more vandalism recorded with names being written on her wheel well doors, a pitot never properly replaced with an adequate model and the substitution of what can only be termed a "clown nose" placed on this prestigious bird. The third attempt at restoration was done after the aircraft was placed in 2 year dry-dock with the rest of the ship for overhaul. This was also for political reasons, as the city wanted the docks refitted and could not do it with Intrepid in place. This situation has occurred for three reasons:

1. Due to the fact that the lender museum had no recourse and the political involvement of the host museum with a money wielding owner and

the military, the lender museum had to resort to undercover operations to protect its aircraft. It was a success at least for a time.

2. After that operation had outlived its welcome, the lender museum was once again being stalled by the host museum and could only rely on some reports of the aircraft's mistreatment. It chose to do nothing and the situation seriously deteriorated. Due to the fact that the lender museum once again was tasked by external reports of mistreatment of the aircraft, the Nat'l Museum of the USAF's director found it necessary to threaten the Intrepid with aircraft removal. if something wasn't done. We must interject that at this point, the Intrepid had no curatorial staff and had not informed the Nat'l Museum of the USAF of its situation which was in violation of its 501C tax code as a museum. This jeopardized not only the Intrepid but the Nat'l Museum of the USAF as well. Since then, it has been rectified.

As can be seen from the Intrepid situation, it had political concerns and pressures by rich owners and a board of directors unfamiliar with the working of a true "A" class museum. This aircraft would have had a safe home in the New England Air Museum. Because of political favors, the aircraft has been left in danger, has deteriorated and still remains in the hands of a museum that only operates when threatened. Another recently added case to this is the acquisition of a Space shuttle on a flight deck under a tent. During Hurricane Sandy, the tent and the Shuttle were damaged. There was no reason for a shuttle to be placed on a flight deck seventeen stories above the Hudson River. Again, it was a point of favoritism in awarding the shuttle to the museum.[1]

[1] See " NASA and the Shuttle Shuffle" Jeannette Remak 2012, Phoenix Aviation Research.

Technical Issues in Caring for A-12 Blackbirds

Restoration issues:

Restoration Philosophy and Policy

According to Louis Casey, Curator of the National Air and Space Museum in a policy written in 1969, Mr. Casey says:

"During the restoration process, extreme care should be taken to preserve intact, existing material. In making the specimen "like new", we can destroy the research value of the specimen. Try as we may it is difficult if not impossible to restore a specimen to its former or original condition. The general tendency for laymen to "restore" vintage aircraft to a like new condition should be resisted at all costs. As a national museum we should expend the time and energy necessary to preserve the original materials and details.

There has to be a first time for process and preservation. We should intensify our efforts in that direction as hopefully other museums will look to us, NASM for guidance. We should prepare ourselves for that challenge and responsibility."

The question of "New Look" over Restoration gives an unquestionable new look with no evidence of aging or appearance of ever being used. Restoring or preserving originality retains evidence of usage and aging while returning aircraft to operational status when being taken off line. That is what is to be displayed, Pristine is not operational and takes away from the history of the aircraft.

Terminology

Original

A specimen that can be shown to be in the original as built configuration or as a modified by the user, that remains unaltered from the time it ended operational status.

Restored Original

As artifact composed of at least 50% original components (by surface area or volume) and the remainder returned to accurate early condition made with the same materials, components and accessories.

Reproduction A reasonable facsimile in appearance and construction of an aircraft made with similar materials and having substantially the same type engine and operating system.

Preservation Act of sustaining and maintaining cultural and natural resources that have been identified as significant and or threatened and that warrant protection.

Conservation As technology of preservation, conservation is the scientific investigation of materials environment and with those things responsible for deterioration of cultural resources.

Preservations and Conservation are similar and designed to keep an object in its existing condition. Restoration: is more an active process which attempts to undo changes cause by past deteriorations.

The professional view of restoration is to document the aircraft in the collection in its time period they were built. The Aircraft should not only be preserved but exhibit the technology of their times. Real reasons for compromise in quality is not cost or time, but poor leadership by the curator attitude, lack of knowledge and weakness in management of the team leader.

The main objectives in restoration are to return the object to earlier appearance or working conditions Restoration and conservation are not but not necessarily the same. Restoration treatment presents a greater risk to the integrity of an object than conservation does. In restoration it is often desirable to replace damaged parts, refine entire surfaces or eliminate later modifications. Restoration treatments are more extensive and intrusive. Risks include misinterpretation of physical evidence and removal of historically significant modifications or materials.

Corrosion control Cleaning agents, preservatives such as coating, electrolytic phase inhibitors and waxes can be used and decisions on how to best use them are important to the process.

Restoration Techniques

In aircraft restoration the intent should be preserve originality regardless of the amount or operational wear that show on the aircraft. It takes more time to repair a damaged part than to replace it.

Signs of active corrosion, albeit slow, must be addressed during the aircraft restoration. Every process must be done to make the aircraft last indefinitely not just for normal manufacturer lifetime. As technicians are working on these aircraft, it must be emphasized that there should be a bond between the tech and his work, and not just treating it like another job. This can't be stressed enough.

LEGACY

There is something else to be added here and it is of great importance, something that is overlooked in many cases. It is called "LEGACY" . While we are looking at all the technical aspects of what you need to do to keep an Aviation museum up to snuff, but there is something else to address.

The men and women who worked in these "Black Programs", have given their lives to these aircraft and all they stand for. They are known as "Silent Warriors". They served in silence, many at Lockheed, some at Area 51, others in tankers service in the USAF to keep these blackbirds, the A-12s and the SR-12s, the MD-21s and the YF-12s fed while in the air, others in jobs that we may never know about. All of these silent warriors are owed our deepest gratitude and respect for their service to our country.

While doing all the everyday jobs that created these magnificent aircraft, they lived under stresses that still go one today with their successors. Their ***LEGACY*** is **that** A-12 or SR-71 or MD-21 or YF-12 that is sitting in a host museum.

How would you feel, if after leaving a job that you gave your life to, you came back to see the object of that dedication left to ruins, uncared for, and in most cases left to rot.

It hurts and it hurts deep. As this author has seen on many a HABU/SR-71 site posted in Facebook, they feel their work has been ***" stuck on a pole and left to rot"***. That is not right, nor is it fair. It is disgusting and insulting. We could add other adjectives but we will leave it at that, for now. LEGACY is what keeps museums ALIVE!!!! You would not see a B-17 or a B-24, possibly an F-106 or the XB-70 Valkyrie left to wrack and ruin. No, you would not. Why should it be any different for an A-12 or an SR-71?

An aircraft with the stature of both the A-12 and the SR-71 deserve to be treated with the utmost respect because of what they DO SIGNIFY! And that is what all of this is about..... HISTORY... LEGACY .... RESPECT and THANKS FOR YOUR SERVICE.

When you look at an aircraft at a museum, think about the hands that built her, the minds and bodies that flew her, the hands that lovingly cared for these aircraft when in service.

That is a point many museums lose sight of in the day to day battle to keep the doors open. REMEMBER the hands and minds that created, serviced, flew and loved their jobs and their aircraft. Maybe with that thought in mind, we won't see so many A-12s and SR-71s in such poor shape. Museums.... You are responsible for not only the history but the legacy of every aircraft you hold in your museum. There are lives behind every, single aircraft in the collection. They should ALL be treated with the respect due them.

Different Categories of Collection and Restoration Decisions

Category I

Aircraft which are historically significant, by virtue of taking part in some historic event are in Category I. Aircraft in this category will be restored to represent the event for which they are most famous.

Category II

Aircraft which are technologically significant due to a feature or advance development . Aircraft in this category should be restored to represent the event for which it became famous. Technically significant aircraft produced for experimental flight, cabin pressure etc can be classified in this category.

Category III

Aircraft not significant in their own right, but which represent their most widely known operational role.

Marking and configuration historically known to be on the aircraft are preferred. Another issue to be raised for action by a curator's staff is choice of military markings for a CAT III aircraft in the collection as representative types, not in themselves historical aircraft.

In the case of military markings, it is necessary to select a set of markings which designate the theater of operations in which the aircraft was used. The main objective should be to display the aircraft in markings which represent its major role. It is best NOT to use markings of famous individuals or tail numbers of famous aircraft on a general aircraft, but to make a more general identification. Careful research must be done to assure authenticity of markings chosen.

Developing Restoration Configuration Plans

The full history and significance of the aircraft must be documented and understood beyond hearsay and legend that may affect the work plan.
A. What secrets might this aircraft hold that should be looked into?
B . What maintenance problems will this aircraft encounter for safe guarding this aircraft after being restored e.g. Scuff marks, Tire inflation, and paint corrosion.

C. What effect, if any, does restoration have upon enhancing or degrading the aircraft's historical or technological significance?

D. Will the aircraft be at some time the primary subject of an exhibit or will it support an element that relates to a group , since it is new or has gone through a complete restoration with preservation safe guards.

E. What time period will it represent?

F. Do the original and quite mundane markings on the aircraft permit the museum to explain or interpret a facet of its mission, which would not otherwise be possible?

G. In what way does the artifact add educational value or strength to the collection? What technical details need special attention for this function? I.e. cutaways, exhibit graphics.

Three Levels of Exhibit and Preservation

Level I An aircraft including the engine is considered in "pristine condition" and ready for exhibit if it is in stable condition (no deterioration), but the exhibit needs some cleaning, markings changed to return it to its significant role.

Level II A *deteriorated or unstable condition*: These aircraft are not suitable for exhibit without a complete or near complete restoration. The condition may have resulted from heavy usage, or poor care. This condition is often influenced by neglect and even abuse within the museum itself. These aircraft are seldom in stable condition and are fragile, brittle, torn and incomplete. The reason the aircraft was given to the museum is because it was no longer serviceable.

Level III The aircraft was derelict or destroyed. It may be possible to conserve them in present condition. Some museums have used them in dioramas to show how the aircraft was discovered.

In the three levels of conditions, it is obvious that there is a wide range in appearance as well as preservation or stability of museum aircraft between the two and three levels. (i.e. Those in good condition as opposed to those deteriorated or destroyed).

If an aircraft is in a pre-failure state and placed in a museum, without preservation or maintenance it will begin to deteriorate to Level III (failure and heavy restoration needed). There is a wide variation of "restoration" of an aircraft really means: *All necessary action to preserve structure indefinitely ,not only the exterior but the interior and to preserve the technology of the design.*

To add "MAKE UP"(yes, in the sense of touch ups in paint or markings) to an aircraft destroys a museum's integrity and confuses the observer as to what is authentic about the aircraft and what is cosmetic. 'Hollywooding" is ok for aircraft flown or used for movie purposes but not for an authentic museum artifact and should be avoided at all costs!

Funding a restoration project

The meat and potatoes of any restoration enterprise is the funding for the project. As already discussed, if you don't have the bucks, you shouldn't take the aircraft. There has to be a sufficient amount of funds set aside after the escrow payment, if you are taking an aircraft from a museum like the National Museum of the USAF. That will support the restoration after the aircraft comes home to its new museum.

This also requires a place that the aircraft can be worked on in comfort for both staff and the aircraft, preferably indoors. The amounts can vary from a small restoration fund of $25,000 to a larger one of $100,000. This is all contingent on the project that is being brought home.

Many times when an aircraft is brought in, it is cleaned up and put on the exhibit floor and not thought about again until the first few nuts and bolts fall off. Finding out why those nuts and bolts fell off usually leads to a more extensive problem and before you know it, the aircraft isn't looking quite as well as the day it arrived. It pays to have funds and a maintenance schedule on hand from the onset to make sure for every eventuality, and is usually required by the loan agreement

Do No Harm: Physical Restoration of an A-12/SR-71, a delicate process

As with any artifact the first rule is to do no harm. If there is no guideline or basic plan, research it till something is found. Curatorial guidelines should be followed and are the logical place to start:
The Curatorial Guidelines document becomes the baseline of reference that all work on the project must adhere to. It explains objectives to be gotten when the work is complete.

Identification: should list the name of the aircraft, its identification number in the case of the A-12, its article number. It should also contain the Museum accession number and the catalog number.

Significance of the type of aircraft, magnitude of the overall use, its primary purpose, and background of the specific airframe, date of manufacture, is a primary concern that must be researched and confirmed for authenticity.

In the case of the A-12, that can sometimes be found on interior longerons or other parts of the aircraft like the chines which can be removed and where you will see a part number. Again, in the case of the A-12 you will not see Lockheed stamped anywhere on the aircraft. Only research will give you that answer as to whether the part is authentic.

Donor of the aircraft to the HOST Museum.

In the case of the A-12 or the SR-71, the National Museum of the USAF.is the loaner of the aircraft. From its records you can find out the date the aircraft was received, total flying hours which can be found in books on the program for that particular Article number and other information. Hopefully, a summary of previous restorations, preservations, treatments and storage condition would also be included.

Current condition In the case of the A-12, the final configuration is the only configuration.

Fuselage With the A-12 and the SR-71 this needs to be gone over inch by inch due to the problems the A-1/SR-71 can manifest.

Cockpit With the A-12, it is sometimes in good condition, if it hasn't been vandalized. On most A-12s there may be a couple of gauges missing in the cockpit. Some of those gauges were top secret and were removed when the aircraft went out of service.

Wings and Landing Gear In the case of the A-12/SR-71 the tires should not be flat. If not new, they may be painted silver after being blown up with foam so they don't flatten. Engine: No A-12 comes with engines.

Colors and markings The Nat'l Museum of the USAF has the markings for the A-12 and the standard color is black, although there are other configurations like black and silver which was a test flight configuration and not a safe outdoor display configuration.

Concluding comments This should be included in the survey you just completed and allows for notes regarding the A-12's history, also there are any dings, broken parts etc., and also when and if any work had been done.

Titanium issues

As discussed the A-12 is 93% titanium alloy. When cared for properly and treated with the correct materials, the A-12 skin should have no real issues. However, when the skin is exposed to hostile conditions, acid rain, snow, ice, cold, heat, humidity and left unprotected outside to face the elements, the skin will not fair well. Considering the relative age of the A-12 to be at least 40 + years old that may not seem like much in aircraft years. However, with a titanium alloy skin that is treated harshly many things can occur:

A. ***Crazing*** On A-12 skin crazing occurs when the metal is starting to fail. It presents at the rivet hole where you can see fine lines around the flush cherry rivets. This usually means that the titanium is brittle. If it spreads, you will see it in the titanium plates as horizontal and vertical lines around the cherry rivets.

B. ***Kadena Krud*** This is a bright neon green krud that forms at the edges of the skin and at rivet holes. It is caused by high humidity. It was first found when the aircraft was stationed in Okinawa, where the "krud" got its name. It can be treated by washing with distilled water, dried with a soft cotton cloth and rubbing Vaseline around the infected areas.

C. **Saw tooth panels** The saw tooth panels in the wings of the A-12 were used to deflect radar. They are coated in side with radar deflecting material and a sealant compound called RTV. They are also loaded with asbestos. In the case of #122, these panels have been removed cleaned and replaced with new panels and sealed with new RTV to prevent any allergic reaction to aluminum and titanium alloy.

D. ***In the case of tools*** NO CADMIUM PLATED TOOLS CAN BE USED ON THE A-12, YF-12 or SR-71. The aircraft skin is highly allergic to it and can cause severe damage to the skin.

The rudders on some A-12 are made of composite material not metal. In the case of 122 on the Intrepid, the rudders were sanded. The rudders on 122 are made of composite material. When the crew on Intrepid realized this it was too late, the damage was done.

Painting the A-12

The best way to approach painting the A-12 would be to chemically remove the paint. There is a company by the name of NUTEK that specializes in this process. It is done safely and within OSHA and EPA regulations. There is a list available of all the materials safe to use on the A-12 airframe in the **Appendices**. Sanding of any sort should not be used due to the composite nature of the aircraft. Titanium does act as its own sealant. Scraping or sanding that "sealant" off only allows more corrosion to enter into the metal. Another method that should **NOT** be considered is power-spray washing. First, using any water except distilled water will damage the metal. The use of a power spray to wash the aircraft that has **NOT** been sealed against elements of weather (outside displays) only allows **MORE** water to penetrate the aircraft causing more moisture damage.

Another point to painting the A-12. Painting the aircraft midnight blue is not correct. There was a situation recently at a museum in U.S. ROCKET AND SPACE MUSEUM, Alabama which painted an A-12 midnight blue because, someone had read that while the aircraft appears black on the ground at full stop, when she reaches mach speed, her paint turns blue. **FOR DISPLAY PURPOSES**! ***THE A-12 AND THE SR-71 SHOULD NOT APPEAR IN ANY OTHER COLOR THAN BLACK!*** What happens at Mach speed does not represent what the aircraft actually looks like on display. It is confusing and technically wrong to attempt to paint her any other color than black. It would be, as this particular museum was also guilty of, marking the A-12 with NASA markings, of which the A-12 NEVER appeared!

More Restoration Issues

A. Indoor and outdoor exhibit The best and safest place for an A-12 would be an indoor exhibit, climate controlled and away from the elements. Outdoors is also applicable if the aircraft has been restored and prepared to be able to stand the outdoor elements and it is in a dry, arid environment. That consists of the aircraft having open panels to be sealed as in the case of #122. The aircraft would need to be cleaned regularly with distilled water and mild soap at least every 3-4 months.

B. Weather issues Any aircraft that is left outside will be made to endure tough conditions like snow, rain , winds , hurricanes etc. the aircraft outside would have to be prepared for that and positioned safely to withstand it.

C. #132 and Hurricane Katrina Even in the best conditions inside, nothing can withstand a hurricane like Katrina. It was only because of the dedication of the Battleship Alabama Museum staff that this aircraft was not totaled and trashed.

D. #122 and the flight deck on Intrepid This issue has been raised throughout this directive. In closing, if there ever was a worse place to put an aircraft of 122's stature, we haven't found it yet. This placement was due to political pressure put on both the pentagon and the Nat'l Museum of the USAF. In this case, money spoke louder than common sense and principle. It should never be allowed to happen again.

E. #127 and the Deterioration in Alabama Space and Rocket Museum. Because of the lack of Nat'l Museum of the USAF control, the Alabama museum has been allowed to continue to disguise its A-12 as a SR-71 dressed in NASA colors. The current restoration is again a problem with the aircraft painted blue because of a misinterpreted quote by an "arm chair expert".

Politics and the Museum system

What to avoid

If looking for an aircraft to add to a collection, make sure that you have the money, materials, manpower and storage space available. If possible keep open lines of communication between you and your lender museum. Always be prepared to show the aircraft no matter when or how. In essence, keep the aircraft in as best condition possible. Avoid taking an aircraft for the sake of prestige. If you can't support it, leave it alone. Avoid getting into any political struggles to make sure you do get the aircraft. It only leads to grief in the end. Follow the Lender Museums criteria and don't be afraid to ask for support, they will help where they can.

Protection of Artifacts

Artifacts whether they are guns, uniforms or aircraft are precious pieces of history. They are entrusted into the museum system to make sure they are safe and that their historical value as a part of era is preserved intact. They are not meant to be bartered, bought or politically influenced for the sake of prestige.

Maintaining the very secret History of the A-12 OXCART Program

It has taken years to bring the OXCART program to light. The depth of this program is still finding its place in Cold War history. It can't be subjected to political pressures of museums whose only claim to fame is to use the A-12 as a meal ticket for their monetary benefit. It behooves the remaining members/retirees of the OXCART program to continue to uphold the prestige of the OXCART program and to keep the flame of OXCART true and bright for generations to follow and know the truth.

Long term commitment of having an A-12/SR-71 as a museum exhibit

While it will take money and great support of a curatorial and restoration staff, having the A-12 as a museum exhibit has to be one of the most exciting and truly historical assets to any museum. The great rarity of the A-12 and the marvelous OXCART program history to support it will make a great and honorable exhibit to commemorate the work and sacrifice of the Roadrunners and Program OXCART and of course, the venerable SR-71.

The SR-71

At this point, while we are discussing the A-12 Blackbird, much if not all you see here can also be applied to the magnificent SR-71. They are birds of a feather and need to be treated with the same care as her predecessor. We will not forget that there are SR-71s that are outside undergoing the same tragic lack of care. Any procedures sited here can be used for the SR-71. The SR-71 has a magnificent record of achievement, more speed records broken than there is room for. Missions from Kadena AB in Okinawa, Japan, where she got her nickname "HABU" after a deadly, night stalking snake that prowled the underbrush, to Libya and the list goes on. The men and women who worked on the A-12 and the SR-71 have much to be proud of. They couldn't tell their stories while in service, but they can tell them now. That is why it is imperative that both these magnificent aircraft are kept in prime condition as display to allow future generations to know just what it was to be a “HABU”.

Oxcart’s History-- A Sacred Trust---- The Legacy of the Roadrunners and the OXCART program

Much has been said about the A-12 as an artifact. What needs to be added to that statement is the real reason for the A-12, the OXCART program. The OXCART program was run in the deepest of CIA black programs. OXCART also revolutionized high speed aviation and high altitude reconnaissance with Mach III speed. The odd part about all of this is no one knew anything about it, which is how well kept a secret it was. Now, that the aircraft has surfaced and been placed on exhibition, it deserves to have its place in United States Cold War history and to have that history displayed correctly. This can be accomplished by the vigilance of the retirees of the program and the fact that by their very presence as living history, they can demand that the museums involved treat their history with the respect that it deserves. If there was ever a time to speak to the generations to come, it is now to make sure that the work and sacrifice that the A-12 Blackbird expresses, demands the respect that the OXCART program deserves. This also holds true for the magnificent history of the SR-71. She was in service longer and more visible than the deep, secret A-12. SENIOR CROWN, the SR-71 program has a lot of material written about it. It should be prized as much as the A-12 as an exhibit. But most of all, remember those that built, serviced, flew and lived the A-12 and the SR-71. They deserve the respect for their service and they deserve to see their legacy, these magnificent aircraft displayed beautifully for their viewing and the viewing of others that will come later. It's all about the HISTORY, not politics, not one-upmanship, or anything else..... THIS is our Country's history and should be displayed with honor, dignity and pride.

Appendices

AFT FUSELAGE, WING AND NACELLE ASSEMBLY

LIST OF ILLUSTRATIONS

Changed 1 March 1966 ii

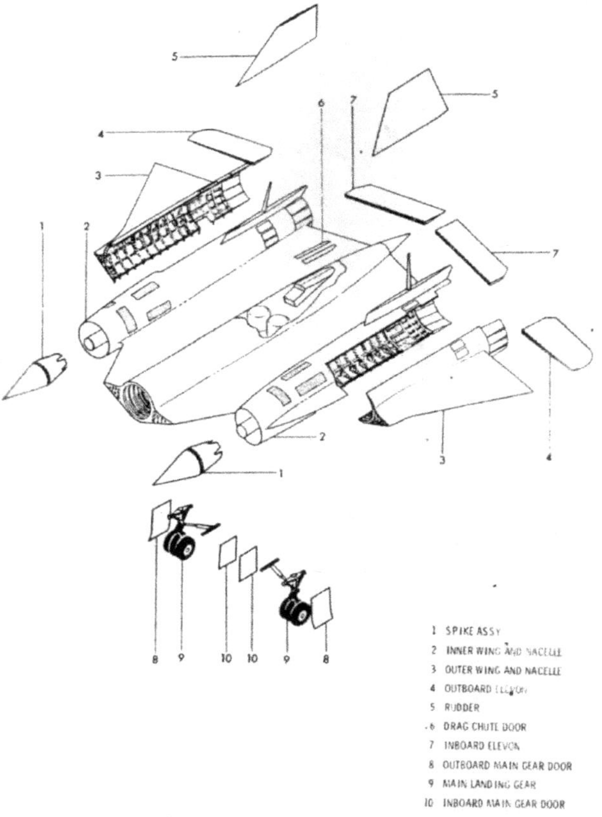
1
2
3
4
5
6
7
5
7
4
3
2
1
8
9
10
10
9
8
1 SPIKE ASSY
2 INNER WING AND NACELLE
3 OUTER WING AND NACELLE
4 OUTBOARD ELEVON
5 RUDDER
6 DRAG CHUTE DOOR
7 INBOARD ELEVON
8 OUTBOARD MAIN GEAR DOOR
9 MAIN LANDING GEAR
10 INBOARD MAIN GEAR DOOR

Figure 5-1. Aft Fuselage (Mid-Body) Wing and Nacelle

5-17. Wing Surface Panels. (See figure 5-5.) The inner wing surface panels are of multiple layer titanium alloy formed sheet construction. The lower surface panels are permanently installed and sealed to the wing structure for fuel retention. The upper surface panels are removable for access to the wing interior. The inside surfaces of the wing panels are formed into corrugations. Outer surfaces of the panels are beaded in the chordwise direction with beads located between the inner surface corrugations. The beaded and corrugated portions of the panels are spotwelded together. This type of construction allows the panels to expand and contract with changes in temperature. Refer to paragraph 5-53 for further details of removable upper surface panels.

5-18. <u>Nacelle Inlet and Inner Nacelle Half.</u> (See figure 5-[illegible].)

5-19. Each engine nacelle is composed of the engine air inlet section, the inboard nacelle half, and the outboard nacelle half. The inlet section and inner nacelle half are built onto the inner wing. The outer nacelle half is built onto the outer wing.

5-20. Nacelle Inlet. The nacelle air inlet is a barrel-section structure extending from the front lip of the nacelle rearward to the front beam of the inner wing aft box section. The inlet barrel section provides mounting and enclosure facilities for the inlet spike and centerbody. (See figure 5-6.) The aft end of the inlet is built directly onto the inboard nacelle half, and structural pin joints connect the inlet section to the outboard side of the inner wing forward box section. The pin joints permit relative motion between the inlet barrel and wing during expansion and contraction from temperature changes. The outboard side of the inlet supports a plastic chine section that fairs into the outboard wing leading edge. Ship serials 121 and 124 differ in that A-110AT metal panels replace the silicone-asbestos panels.

5-21. The inlet centerbody is a strut-mounted assembly located in the center of the inlet barrel section and serves as a mounting facility and actuating component enclosure for the inlet spike. Refer to paragraph 5-[illegible]7 for a general description of the spike. Refer to A12-2-4 Technical Manual for spike maintenance procedures.

5-22. Inner Nacelle Half. The inner nacelle half forms half of the engine compartment and is built as an integral part of the inner wing. Each beam of the wing aft box section is connected to a nacelle ring frame member on each side of the airplane. Nacelle ring frames are spotwelded, built-up, I-section, titanium alloy, frame assemblies except for the forged frame at [illegible] 1032. Longitudinal members of the nacelle half are built-up of formed sheet and extruded titanium alloy beams. These beams support the nacelle hinge fittings at the top edge and support the nacelle connector plates (flapper plates) at the bottom edge.

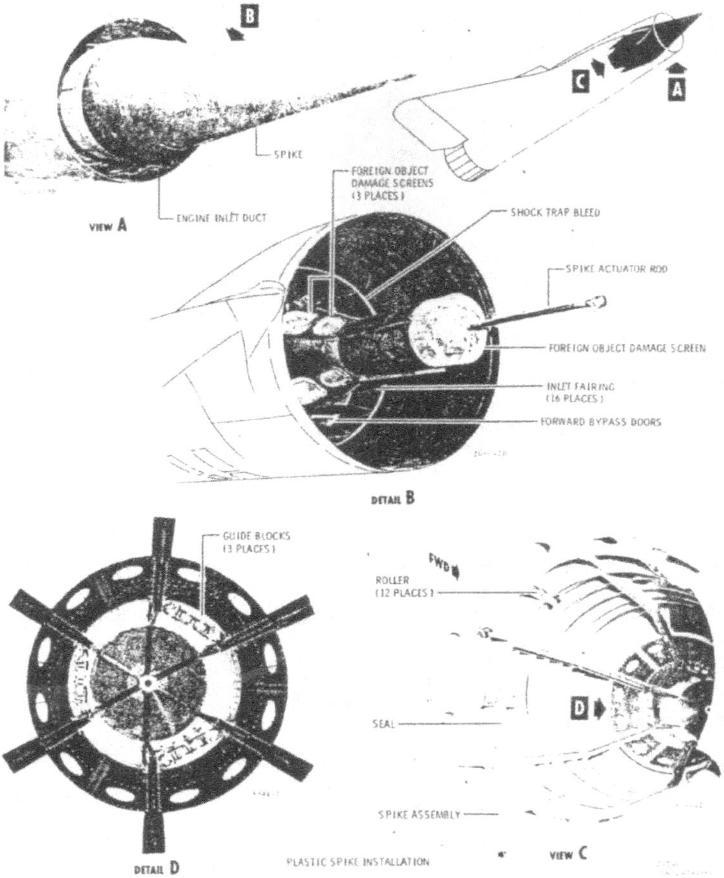
B
C
A
SPIKE
ENGINE INLET DUCT
VIEW A
FOREIGN OBJECT
DAMAGE SCREENS
(3 PLACES)
SHOCK TRAP BLEED
SPIKE ACTUATOR ROD
FOREIGN OBJECT DAMAGE SCREEN
INLET FAIRING
(16 PLACES)
FORWARD BYPASS DOORS
DETAIL B
GUIDE BLOCKS
(3 PLACES)
FWD
ROLLER
(12 PLACES)
D
SEAL
SPIKE ASSEMBLY
DETAIL D
PLASTIC SPIKE INSTALLATION
VIEW C

Figure 5-6. Nacelle Components

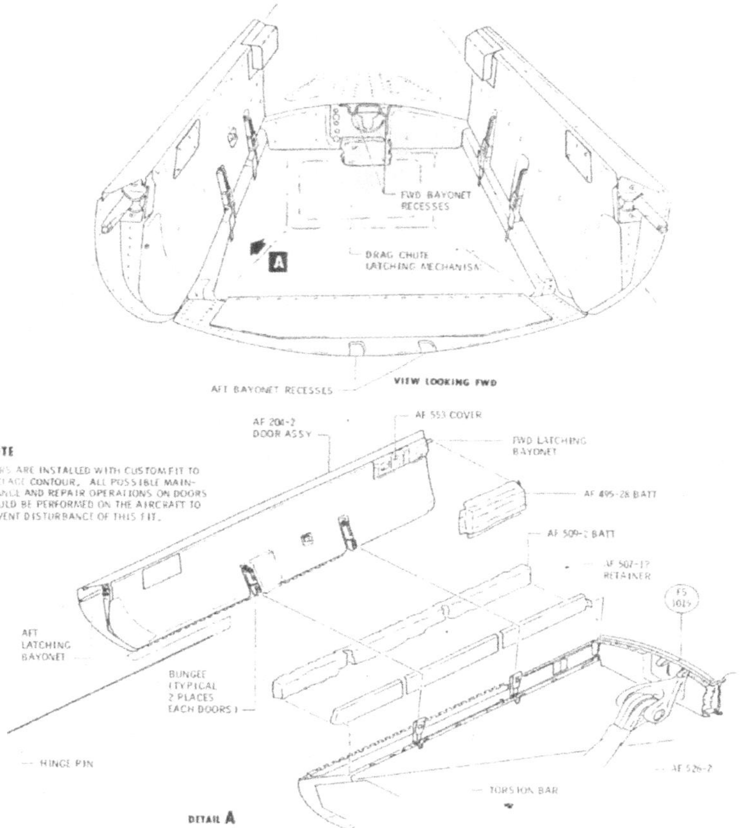
FWD BAYONET RECESSES
A
DRAG CHUTE LATCHING MECHANISM
AFT BAYONET RECESSES
VIEW LOOKING FWD
AF 204-2 DOOR ASSY
AF 553 COVER
FWD LATCHING BAYONET
AF 495-28 BATT
AF 509-2 BATT
RETAINER
FS 1015
AFT LATCHING BAYONET
BUNGEE (TYPICAL 2 PLACES EACH DOORS)
HINGE PIN
AF 526-2
TORSION BAR
DETAIL A

NOTE

DOORS ARE INSTALLED WITH CUSTOM FIT TO FUSELAGE CONTOUR. ALL POSSIBLE MAINTENANCE AND REPAIR OPERATIONS ON DOORS SHOULD BE PERFORMED ON THE AIRCRAFT TO PREVENT DISTURBANCE OF THIS FIT.

Figure 5-8. Drag Chute Doors

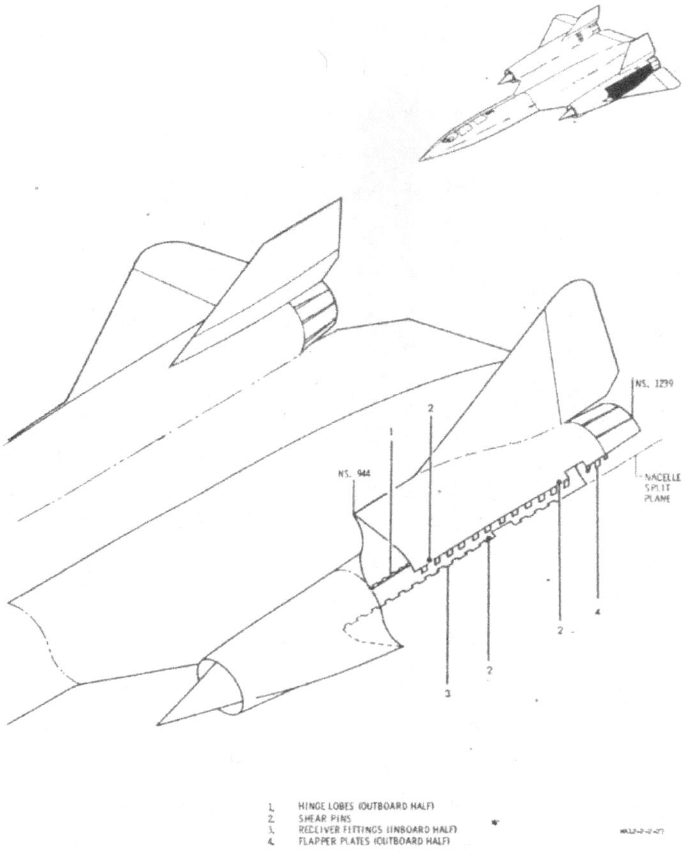
NS. 1239
NS. 944
NACELLE SPLIT PLANE
1
2
2
2
3
4
1. HINGE LOBES (OUTBOARD HALF)
2. SHEAR PINS
3. RECEIVER FITTINGS (INBOARD HALF)
4. FLAPPER PLATES (OUTBOARD HALF)

Figure 5-7. Outer Wing and Nacelle Half

5-36. LAC Serial No. 121, 124, 134 and 135 have rudders made of titanium alloy. LAC Serial No. 122, 123, 125 through 133 have rudders made of plastic materials.

5-37. Metal rudders are built with a structural box section in the center. A leading edge section, trailing edge section and tip are built onto the central box section.

5-38. Plastic rudders have basic frame members of titanium alloy. Subordinate members, including some of the ribs, and spars, and the exterior surface panels are made of bonded silicone asbestos reinforced plastic materials.

5-39. Elevons.

5-40. The elevons are control surfaces hinged at the upper trailing edge of the inboard and outboard wings. They function both as elevators and ailerons in response to movement of the pilot's control stick. At the extreme positions of the stick, maximum elevon travel is 35 degrees upward and 20 degrees downward.

5-41. Construction. The forward section of each elevon is of a structural box, beam rib and skin type titanium alloy construction. The trailing edges are constructed of triangular shaped titanium alloy and silicone asbestos reinforced plastic panels. The triangular plastic panels mesh with the triangular titanium alloy panels. LAC Serials 121 and 124 differ in that metal filler panels, consisting of A110AT titanium skin and that section stiffeners are used in place of the silicone-asbestos panels. Similarly, the silicone-asbestos trailing edge vee is replaced by a sheet metal vee of A-110AT titanium alloy.

5-42. Inboard Elevons. Inboard elevons are attached to six hinge points and six actuating cylinder rods at the inboard wing trailing edge.

5-43. Outboard Elevons. Because outboard wings are not as thick as inboard wings, the elevon actuating cylinders must be smaller; therefore more cylinders are required. Outboard elevons are connected to 17 hinge connections and 14 actuating cylinder rods.

5-44. Tail Cone.

5-45. The tail cone is attached to the fuselage at FS1226 and extends to FS 1310. It consists of typical terminal stringers, rings and skin panel construction. The tail cone houses fuel valves and the control system mixer. Top and bottom doors provide access to plumbing.

NOTE

Although three men can push the rudder to the desired position without hydraulic power, the preferred method is by hydraulic actuation. Hydraulic actuation of flight controls shall be performed only by a qualified flight controls mechanic.

If rudder must be positioned without hydraulic power, the servo followup lever shall be disconnected to prevent damage to the mechanism. This shall be done only by a qualified flight controls specialist.

d. Disconnect gudgeon arm- to-rudder link at gudgeon arm by removing retainer screws and pins (See figure 5-10.)

NOTE

The gudgeon arm is connected to the rudder by a bone-shaped link. The link is attached to the gudgeon arm by two pins. An inner pin is encased by a sleeve-type outer pin. Two 1/4-inch bolts retain the pins in position.

e. Remove rudder post access panel to gain access to rudder post lock nut installation.

f. Remove lock-wire, lock bolt, two washers and lock.nut, and one lock nut washer. (See figure 5-10.) Retain removed parts.

g. Recheck sling tension and carefully hoist the rudder clear of the aircraft.

h. Position the removed rudder on AG 29 rudder cart.

5-66. Rudder Installation. (See figure 5-11.)

a. Prior to hoisting the rudder into place, make sure that the washer, (see 8, figure 5-11; 10, figure 5-12) is in place.

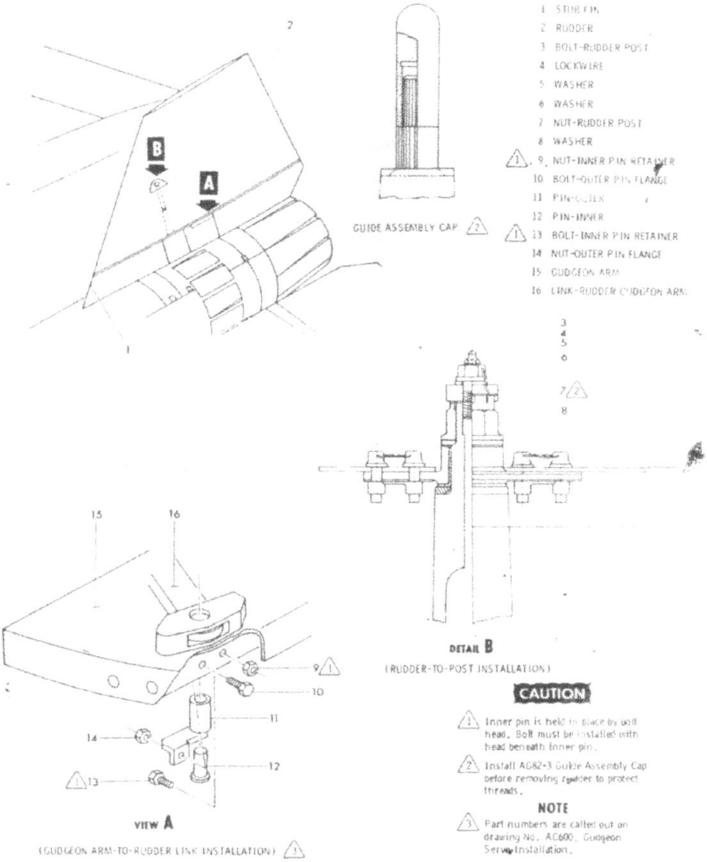
2
B
A
1
GUIDE ASSEMBLY CAP 2
1 STUB FIN
2 RUDDER
3 BOLT-RUDDER POST
4 LOCKWIRE
5 WASHER
6 WASHER
7 NUT-RUDDER POST
8 WASHER
1 9. NUT-INNER PIN RETAINER
10 BOLT-OUTER PIN FLANGE
11 PIN-OUTER
12 PIN-INNER
1 13 BOLT-INNER PIN RETAINER
14 NUT-OUTER PIN FLANGE
15 GUDGEON ARM
16 LINK-RUDDER GUDGEON ARM
3
4
5
6
7 2
8
15
16
9 1
10
11
14
12
1 13
VIEW A
(GUDGEON ARM-TO-RUDDER LINK INSTALLATION) 3
DETAIL B
(RUDDER-TO-POST INSTALLATION)
CAUTION
1 Inner pin is held in place by bolt head. Bolt must be installed with head beneath inner pin.
2 Install AG82-3 Guide Assembly Cap before removing rudder to protect threads.
NOTE
3 Part numbers are called out on drawing No. AC600, Gudgeon Servo Installation.

Figure 5-10. Rudder Removal (Sheet 1 of 2)

5-53. Wing Upper Surface Panels. (See figure 5-2.) Wing upper surface panels are removable to permit access to the interior of the fuel tank and dry bay areas of the wing.

5-54. Panels over fuel tanks are sealed around the edge with a high-temperature silicone rubber gasket material. Mid-bay panel attachment screws are sealed with gasket material bonded to a thin metal washer. Ends of inner surface corrugations of the fuel tank (or wet-bay panels) are capped off with a metal cup and sealed with tank sealant.

5-55. Panels over non fuel tank "dry-bay" areas are not sealed around the edges or at mid-bay attachment points. Ends of inner surface corrugations of these panels are capped off with a modified hat section and are not sealed.

5-56. Wing Fillets.

5-57. Removable wing fillets provide a smooth contour between the fuselage and the upper and lower wing surfaces. The fillets are built-in sections made of titanium alloy sheet and formed parts. There are eight upper fillet sections and six lower fillet sections. Fillets are held in position by interlocking fixtures which allow the sections to expand and contract with temperature changes. Fixtures beneath the fuselage edge of the fillets engage brackets on the fuselage. Sliding pins at the wing edge of the fillets engage slotted fittings on the wings. Except for the section over the wheel well, the trailing edge of each section overlaps the leading edge of the section behind it. This makes it necessary that fillets be removed and installed in proper sequence. (Refer to Removal, Inst. & Maint. in this section.) The upper fillet section over the wheel well serves as a keystone section to cap off the other sections of upper fillets. Screws are used in the installation of this section.

5-58. Open areas under the ends of the fillets at the main wheel well are blocked with triangular shaped metal partitions. These partitions are fastened with screws along one edge and anchored with tape along the other two edges. This allows the partitions to move with the fillet expansion and contraction.

5-67. (Cont'd)

NOTE

If A- dimension is not according to the nominal dimension specified in the delta note on the figure it is necessary to replace the fitting, 9, on figures 5-11 and 5-12. In the left column of the fitting table on the figure, find the range wherein the actual measurement of dimension - A fits. Then remove the rudder and replace the fitting (9, figure 5-11 and 5-12 with the fitting specified in the center column of the fitting table.

If the clearance of dimension - A is within tolerance and the clearance at the front is not within the tolerance shown, consult the manufacturer's engineering representative for proper dispostion.

e. With the rudder clearance within acceptable limits, complete installation of rudder locknut according to the figure 5-11 or 5-12.

CAUTION

Make sure chamfered washers are installed in proper direction according to the figure 5-11 or 5-12.

Carefully observe special torque values given on the figure 5-11 or 5-12.

Lubricate screws and bolts according to the figure 5-11 or 5-12.

f. Reassemble gudgeon arm connection dissembled in paragraph 5-65.

g. Lubricate pin retainer bolts according to requirements specified in Section II of this manual.

5-69. (Cont'd)

h. Remove hinge bolt assemblies. Retain removed parts in suitable container.

i. Move elevon aft to disengage it from the wing trailing edge.

5-70. Elevon Installation. (See figure 5-13.) Elevons being reinstalled may be installed by reversing the removal procedure.

CAUTION

Extreme care must be exercised to prevent interference between fastener assemblies and surrounding structure. Where clearance is critical, lock-wire must be snugged up and shaped around the fastener nut so as to provide not less than 0.020-inch minimum gap at all possible points of contact when the elevon is moved through its full range of up and down travel. Cracked hinge fittings could result if interference develops.

5-71. Lubricate the installation according to A12-2-1 Technical Manual. Lubricate bonding jumper screws according to section II of this manual.

5-72. Elevons being installed for the first time must be checked for clearance throughout full travel of actuator pistons. Refer to A12-2-7, Technical Manual. This shall be approximately 36 degrees up and 21 degrees down. The elevons must clear adjacent structure throughout this travel.

NOTE

Cutouts in the lower leading edge skin may be filed locally to clear cylinders and hinge fittings.

5-73. Wing Upper Surface Panels.

5-74. Panel Removal. (See figure 5-14.) The following precautions must be carefully observed before starting to remove panels.

RIGHT INBOARD WING
HINGE FITTING BOLTS
ACTUATOR ROD BOLT
RIGHT INBOARD ELEVON

RIGHT INBOARD ELEVON CONNECTIONS
(VIEW LOOKING UP AND AFT)

Figure 5-1 . Elevon Removal and Reinstallation (Sheet 2 of 3)

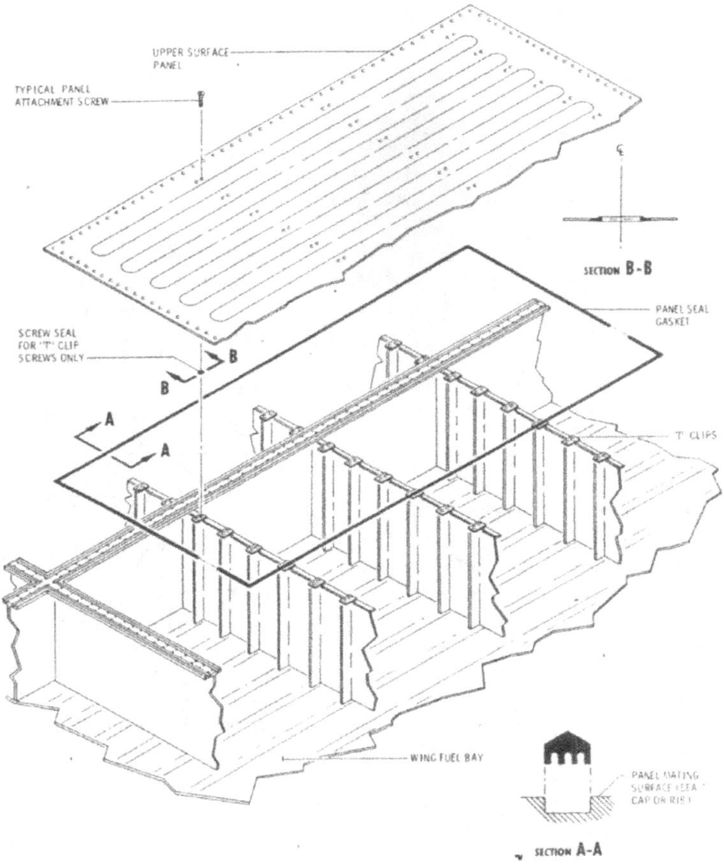
UPPER SURFACE
PANEL
TYPICAL PANEL
ATTACHMENT SCREW
SECTION B-B
PANEL SEAL
GASKET
SCREW SEAL
FOR "T" CLIP
SCREWS ONLY
B
B
A
A
"T" CLIPS
WING FUEL BAY
SECTION A-A

Figure 5-14. Wing Upper Surface Panel Installation

1 AG 56 JACK ⚠1
2 AG57 JACK ⚠2
3 AG 142-3 CONTOUR BOARD ASSY ⚠2
4 AG 5R JACK ⚠1
5 AG 5L JACK ⚠1

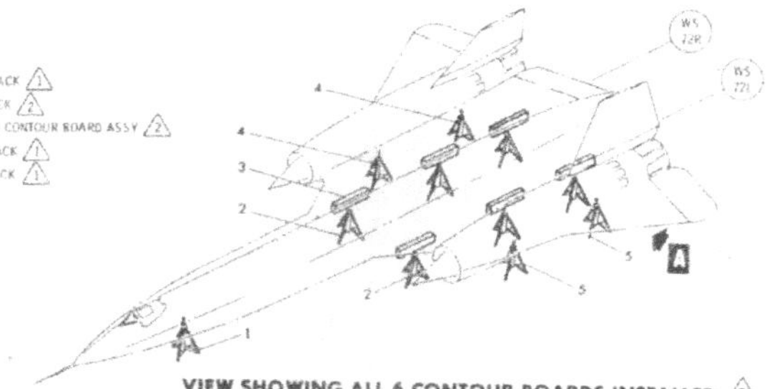
WS 72R
WS 72L
4
4
3
2
2
5
5
1
A

VIEW SHOWING ALL 6 CONTOUR BOARDS INSTALLED ⚠3

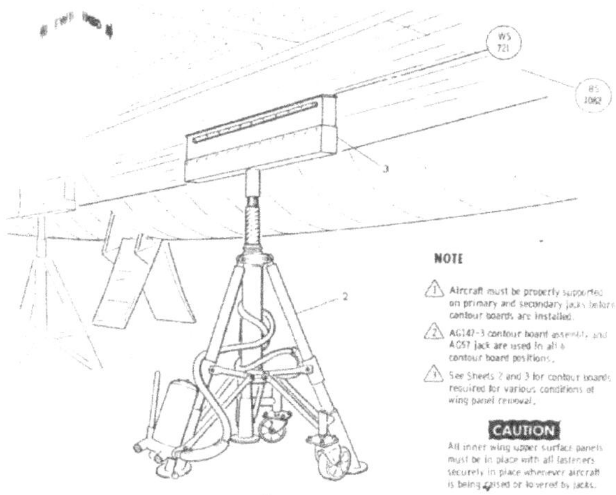
WS 72L
BS 1062
3
2

NOTE

⚠1 Aircraft must be properly supported on primary and secondary jacks before contour boards are installed.

⚠2 AG142-3 contour board assembly and AG57 jack are used in all 6 contour board positions.

⚠3 See Sheets 2 and 3 for contour boards required for various conditions of wing panel removal.

CAUTION

All inner wing upper surface panels must be in place with all fasteners securely in place whenever aircraft is being raised or lowered by jacks.

VIEW A

Figure 5-15. Wing Support for Removal of Upper Surface Panels (Sheet 1 of 2)

5-74. (Cont'd)

NOTE

If seals have not had flight exposure they may be reused if not damaged. These seals must be thoroughly cleaned and subjected to 100 percent inspection before reuse.

f. Carefully scrape old cement from edge gasket groove and tee-clip pads.

WARNING

Extreme care must be exercised in using solvents around tank sealed areas. Methyl ethyl ketone (MEK) and methyl isobutyl ketone (MIBK) soften viton panel gaskets and tee clip circular seals. These solvents are highly flammable and care must be taken to shield all nearby electrical equipment. Proper ventilation must be provided to protect personnel.

CAUTION

Extreme care must be taken when using solvent around existing tank sealed areas to protect existing sealant from solvent action.

If viton gaskets or seals must be cleaned, use a rag moistened with toluene as necessary.

5-75. Panel Installation. (See figure 5-14.) Panel attachment screws must be of the proper length. Screws that are too long damage dome nuts or fail to tighten securely. Screws that are too short may not engage the nut locking device.

a. Check dome nuts for damaged threads.

5-75. (Cont'd)

NOTE

If holes do not line up, jockey nacelle jacks and wing contour boards as necessary.

e. Torque wing panel attachment screws to the following values:

SCREW SIZE	TORQUE
10/32	40 pound-inches
1/4	90 pound-inches
5/16	200 pound-inches

CAUTION

Make sure that no two adjacent screws are tightened consecutively. If screws are tightened consecutively, the panel gasket will be forced ahead and extrude out of the groove.

Tighten screws in a back and forth pattern to ensure that the panel is drawn down uniformly.

f. Pressure check tanks according to instructions in A12-3 Structural Repair Manual.

g. Retorque all screws in the reinstalled panel at each of the following 5 times:

After aircraft is removed from jacks

After first engine run up following panel installation

After each of the first three flights following panel installation

5-76. Wing Fillets.

5-77. Removal and Installation Sequence. (See figure 5-16.) Upper and lower wing fillets must be removed and installed in proper sequence. This sequence is presented on the figure 5-16.

5-78. Silicone Asbestos Laminated Parts.

5-79. Drying of Fuel-Socked Plastic Parts. Fuel soaking is detrimental to plastic parts. Although not mandatory for safe flight, good maintenance practice dictates that soaked panels be thoroughly dried as soon as practical after soaking, thus prevent possible blister damage.

5-80. A fuel-soaked part is defined as a part which has been exposed to fuel flow for a period of 3 hours accumulated time between flights which attain speeds above Mach 2.3.

5-81. Plastic parts which have been fuel-soaked can be dried out satisfactorily by either of the following methods:

METHOD A: This method of drying is to be employed when time between flights permits its use. This method consists of removing part from airplane and drying in an oven (AG167) at 300° to 350°F for a minimum of 4.0 hours. If sufficient time is available the optimum oven time is 24 hours.

METHOD B: This method of drying is to be employed when it is not considered practical to use Method A. Method B consists of the application of hot air blown directly on the fuel-soaked area of the part. Hot air is to be obtained from an AG1387 heater and 4AG1487 high temperature duct assembly. The plastic surface temperature must be maintained at 300° to 350°F for as long as airplane down time will permit up to a maximum of 24.0 hours. The temperature of the surface must be monitored by use of thermocouples or thermometers placed in the center of hot air impingment. The temperature measuring device can be anchored in place by use of Mystic 7000 G tape.

WARNING

If surface temperature is allowed to exceed 400°F prior to complete drying, blister damage may occur. Blister damage must be repaired by methods presented in A12-3 Structural Repair Technical Manual.

3-5. Instruments:

a. Remove protective covers.

b. Carefully inspect instruments for evidence of deterioration. Remove and replace as necessary.

c. Conduct thorough checkout of pitot-static system.

3-6. Landing Gear:

a. Remove preservative from actuators and struts if not previously accomplished.

b. Service tires as per A12-2-3 T.M.

c. Conduct checkouts as required.

3-7. Brake System:

a. Flush/bleed system.

b. Service as per A12-2-3 T.M.

c. Conduct anti-skid and operational checkouts as per A12-2-3 T.M.

3-8. Hydraulic Systems:

a. Flush all hydraulic systems thoroughly.

b. Service systems as per A12-2-3 T.M.

c. Conduct cold leak check.

d. Conduct hot leak check.

3-13. Cockpit:

a. Replace loose equipment.

b. Replace ejection seat pyrotechnics.

c. Conduct appropriate checkout/inspection.

3-14. Inspection Requirements:

NOTE

The inspection requirements for processing the airplane into service following storage will vary according to the Duration of Storage, ambient conditions at storage site, amount of service bulletin incorporation required, etc.

a. Giving full consideration to the specific storage conditions of the airplane being returned to service, conduct the applicable inspections in Sections VI, V, and IV of the current A12-6 Inspection Manual.

b. Conduct BASIC POSTFLIGHT inspections (Section II) of the A12-6 T. M.

c. Conduct PREFLIGHT inspections of Section I of the A12-6 T. M. Conduct Anti-FOD and engine performance runs.

d. Perform a Functional Engineering Test Flight.

AIRFRAME MATERIALS

LIST OF ILLUSTRATIONS

SECTION III

AIRFRAME MATERIALS

3-1. AIRFRAME MATERIALS.

3-2. General. The materials used to withstand the high temperatures imposed by the operating environment of the airplane are titanium alloys, corrosion resistant steel, nickel alloys and silicone-asbestos laminates.

3-3. Titanium. Titanium is characteristically light, strong, heat-resistant, and non-magnetic. Its strength compares closely with corrosion resistant steel while its nominal density is but slightly over half that of corrosion resistant steel. The titanium alloys mostly used in the aircraft are designated as A-A110AT (5Al - 2.5Sn), B-120VCA (13V - 11Cr - 3Al), and C-120AV (6Al - 4V). A-110AT contains approximately 5 percent aluminum and 2.5 percent tin. B-120VCA contains approximately 13 percent vanadium, 11 percent chromium and 3 percent aluminum. C-120AV contains approximately 6 percent aluminum and 4 percent vanadium.

3-4. Corrosion Resistant Steel. The basic corrosion resistant steel alloy for high temperature applications is designated A-286. This is a heat treatable alloy that contains approximately 15 percent chromium, 26 percent nickel, 1 percent molybdenum, and 2 percent titanium as main alloying agents. This alloy will withstand temperatures of approximately 1200°F.

3-5. Nickel Alloys. Two nickel alloys, Rene' 41 and Hastelloy "X" are used on the aircraft in areas subject to extremely high temperatures such as the engine nacelle ejector section. Rene' 41 is a nickel base metal alloyed with chromium, iron, molybdenum, cobalt, titanium and aluminum. It withstands temperatures up to approximately 1600°F. Hastelloy "X" is a nickel base metal alloyed with chromium, iron, and molybdenum. It withstands temperatures up to approximately 2200°F.

3-6. Silicone - Asbestos Laminates. Silicone - Asbestos laminates are used extensively in areas which operate in the 400°F to 750°F temperature range. In wing and fuselage areas this material is installed in the form of replaceable panels.

3-7. Material Protection. The temperatures and stresses imposed upon aircraft materials in supersonic flight make it necessary to exercise greater care and damage prevention measures than is required for subsonic flight. Particularly important is the prevention of stress corrosion cracking, notches and surface cracking. Care in handling and assurance that only compatible materials are used in contact with the aircraft are prime factors in protecting the aircraft structure.

3-8. Stress Corrosion Cracking. (See figure 3-1.) An important item to protect the structure against is stress corrosion cracking. Stress cracking and/or severe corrosion can be caused by intermetallic reactions with certain materials such as mercury, mercury amalgams and cadmium. If the aircraft materials were allowed to become contaminated with such elements and then exposed to heat, the contaminated area may be subject to cracking as a result of the contamination, high temperature and stresses imposed. Certain compounds containing members of the halogen family (chlorine, bromine, iodine, fluorine) decompose into acids when subject to heat. Materials containing the halogens, chlorine, bromine, or iodine shall not be used in contact with titanium alloys unless they appear on the list of compatible materials provided in figure 3-3. Teflon and Viton rubbers (stabilized) are compounds containing fluoride but are stable enough to be used safely.

3-9. Cadmium Precautions. Cadmium plate is not used in any area of the airplane which is subject to temperatures of 450°F or higher. Refer to paragraph 3-8 and see figure 3-1 for information on cadmium as a cause of stress corrosion cracking.

3-10. Purchased equipment containing cadmium plated parts are limited to areas not subject to temperatures over 450°F and limited to items specifically authorized by engineering drawing or other written document prepared by the airplane manufacturer.

3-11. All hand tools or other equipment that come into direct contact with the aircraft shall be free of cadmium plate.

NOTE

Chromium or nickel plate are acceptable platings for corrosion protection.

CRACKING OF TITANIUM CAUSED
BY CONTACT WITH CADMIUM PLATE

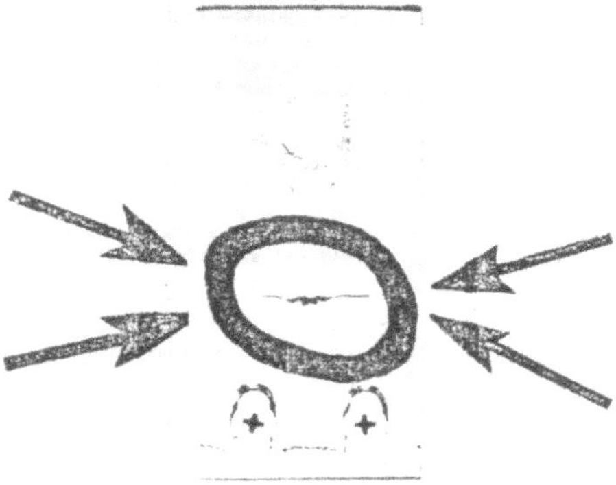

THIS WAS A STATIC TEST OF A TYPICAL SCREW JOINT IN A-110 TITANIUM SHEET. THE CRACK OCCURRED AFTER ONLY A FEW MINUTES EXPOSURE AT 900° F. A CADMIUM-PLATED NUT HAD BEEN USED TO HOLD ON A PIECE OF TEST EQUIPMENT.

3-12. All hand tools that have become contaminated by performing work on cadmium plated parts shall be decontaminated before they are used on any airplane part that is subject to temperatures of 450° F or higher. Decontaminate tools by the following method.

a. Soak contaminated items in cadmium test solution No. 1 for one hour. Paragraph 3-15 presents mixing instructions for cadmium test solutions. Occasionally agitate the solution during the soak.

b. Rinse thoroughly with running water.

c. Perform test for cadmium with test solutions No. 1 and No. 2. (Refer to paragraphs 3-13, 14 and 15.) (See figure 3-2.)

3-13. Test For Cadmium. (See figure 3-2.) Since cadmium is an element commonly used as a plating on tools and hardware, it is sometimes necessary to spot-test an item in question to determine if cadmium is present. The spot-test employs the use of two solutions. Solution No. 1 is ammonium nitrate and water. Solution No. 2 is sodium sulfide and water.

3-14. Procedure. A drop of solution No. 1 is placed on the plated surface and allowed to stand for approximately 30 seconds. A drop of solution No. 2 is added. A yellow color indicates the presence of cadmium. If cadmium is present, steps must be taken to have the parts stripped and replated with a material which is compatible with titanium alloys. Nickel or zinc (cadmium free) platings have been found to be compatible platings.

NOTE

Solutions lose strength in time. The solutions shall be periodically checked with known cadmium-plated parts.

3-15. Preparation of Cadmium Test Solutions. Cadmium test solutions can be prepared by mixing the following ingredients.

SOLUTION NO. 1

6 approximately level teaspoons ammonium nitrate.

6 drops ammonium hydroxide.

8 oz. distilled water.

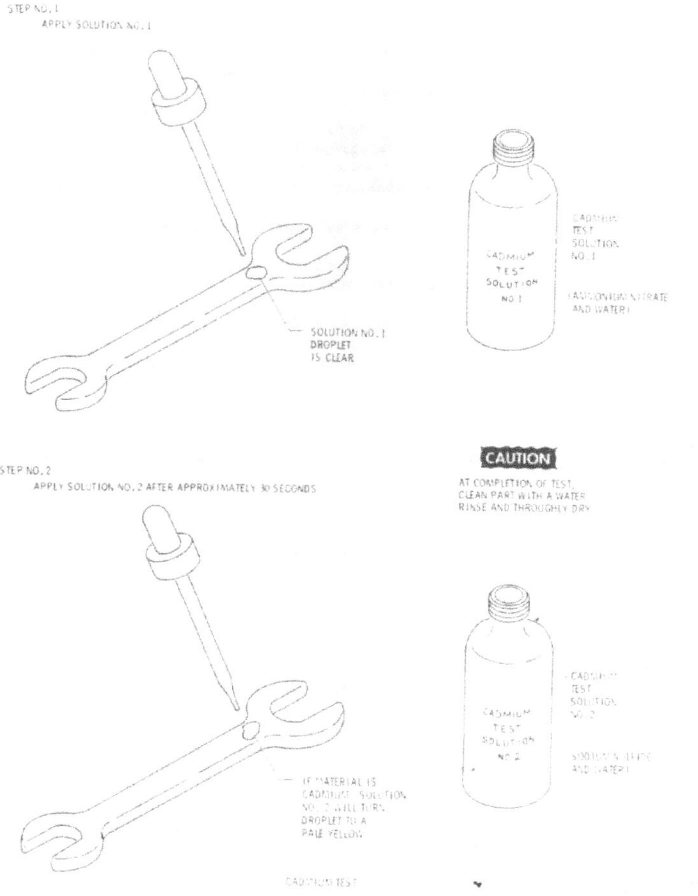
STEP NO. 1
APPLY SOLUTION NO. 1
CADMIUM TEST SOLUTION NO. 1
SOLUTION NO. 1 DROPLET IS CLEAR
STEP NO. 2
APPLY SOLUTION NO. 2 AFTER APPROXIMATELY 30 SECONDS
CAUTION
AT COMPLETION OF TEST, CLEAN PART WITH A WATER RINSE AND THROUGHLY DRY
CADMIUM TEST SOLUTION NO. 2
DROPLET TO A PALE YELLOW
CADMIUM TEST

Figure -2. Cadmium Test Procedure

3-15. (Cont'd)

SOLUTION NO. 2

6 approximately level teaspoons sodium sulfide.

8 oz. distilled water

WARNING

Sodium sulfide is toxic. Avoid prolonged physical contact. If skin is exposed to substance, wash with water.

3-16. Only items of equipment which actually contact the airplane or which need frequent adjustment while being used on or near the airplane need be free of cadmium plate.

3-17. Equipment not used in contact with the airplane shall have all exposed cadmium plated bolts and parts painted with epoxy paint or clear lacquer. This will minimize transfer of the cadmium plate.

3-18. Edge Notch and Surface Cracking. The aircraft materials are subject to extensive expansion and contraction from changes in temperature. Under such conditions the titanium alloys exhibit a readiness to develop cracks from edge notches, surface scratches and gouges. It is therefore necessary to exercise great care and diligence in the prevention of such damage. Specific preventive maintenance measures such as walkway protection and requirements for protective covers are provided in A12-2-1 Technical Manual. Any scratch which can be felt with the thumbnail is sufficient damage to warrant remedial action. Several methods of dressing out and repairing such damage have been approved. Negligible damage limits and repair procedures are provided in A12-3 Technical Manual.

3-19. Compatible Materials. (See figure 3-3.) All materials having any possibility of being used in contact with the aircraft must be tested prior to use to determine compatibility with titanium alloys at high temperatures. This includes such materials as solvents, adhesives, tapes, paints, marking materials, plastics, fire extinguishing agents, and other items which in the past have been commonly used in aircraft maintenance. Such products are continually being tested by the aircraft manufacturer. Those products upon which tests have been completed and found compatible with titanium alloys at high temperatures and stress are listed in figure 3-3. If any item in question is not found in the list of compatible materials, it shall not be used until it has been tested, found compatible, and added to the list in figure 3-3.

CAUTION

Just because an item appears on the list of compatible materials does not mean that it may be substituted for another similar material which is specified on an engineering drawing or in a technical order. In some cases a substitute material may be compatible with titanium alloys but incompatible with other materials such as plastics or sealants. Before attempting a substitution for any specified material, approval must be obtained from the aircraft manufacturer's engineering representative.

3-19A. DECONTAMINATION OF TITANIUM MATERIALS EXPOSED TO CADMIUM. The surfaces and holes of titanium parts that have been contaminated by contact with cadmium can be decontaminated by the proper use of a decontaminating agent as specified in the following paragraphs.

3-19B. Cadmium Stripping Agent. The cadmium stripping agent is a jell that consists of the No. 1 solution from the cadmium test kit mixed with SANTOCEL "C", powder. SANTOCEL "C" (Monsanto Chemical Co.) is available through the S. L. Abbott Co. 4255 East District Ave., Vernon, California.

3-19C. Cadmium stripping jell can be prepared as follows:

a. Prepare the No. 1 cadmium test solution.

6 approximately level teaspoons ammonium nitrate.

6 drops ammonium hydroxide.

8 oz. distilled water.

b. Mix the No. cadmium test solution with SANTOCEL "C" to the consistency required to hold the mixture on the contaminated material.

3-19D. Cadmium Stripping Procedure. Decontaminate titanium parts exposed to cadmium as follows:

a. Apply cadmium stripping jell to contaminated area of part and let set from one to two hours.

NOTE

Apply jell to flat surfaces with a clean, dry cloth or bristle brush. Apply jell into holes with a clean cotton swab on wooden stick.

b. Wipe area clean with clean, dry cloth. Wipe holes with clean cotton swab.

c. After area has been wiped clean with dry wiping materials, wipe again with clean cloth (holes with clean swab) moistened with distilled water.

CAUTION

Avoid flowing water onto area to remove cadmium stripping jell. Flowing water will wash contaminants into uncontaminated areas.

d. Perform test for presence of cadmium using clean white cloths and swabs dampened with solutions No. 1 and No. 2 (Refer to paragraph 3-15.)

CAUTION

Do not flow the liquid test solution onto decontamination area. Flowing liquids could wash contaminants onto uncontaminated areas.

NOTE

A yellow color on the dampened cloth or swab indicates that cadmium is still present. If so, repeat steps a through d.

3-20. (Cont'd)

p. SOLVENTS (Sheet 12)

q. TAPES (Sheet 12)

r. TEXTILES AND FABRICS (Sheet 13)

s. WRITING AND MARKING MATERIALS (Sheet 13)

t. MISCELLANEOUS CHEMICALS AND COMPOUNDS (Sheet 14)

ABRASIVES

Abrasive Wheel - Carborundum A80HFFXZ Tool Mart, Van Nuys, Calif.

Abrasive Wheel - Norton A036TNA Behr-Manning Co. Los Angeles, Calif.

Abrasive Wheel - Norton A60CRAZ Behr-Manning Co. Los Angeles, Calif.

Abrasive Wheel - Norton A30TNAY 143 Reinforced, Black, Behr-Manning Co., Los Angeles, Calif.

Abrasive Wheel - Norton A30TNAY 143 DS Reinforced, Blue, Behr-Manning Co., Los Angeles, Calif.

Abrasive Wheel -Polybond 150 Extra Hard, 3600 Davenport Abrasive Corp., Rockland, Mass.

Abrasive Wheel - Tyco Brand Unitized Wheel, Code 71A Grade-Med. 3M Co., St. Paul, Minn.

Alumina Oxide Plus Water

Carborundum PC-8A Plus Water

Cloth, X wt., Waterproof Silicon Carbide, 3M WET OR DRY, TRI-M-ITE

Crocus Cloth, 3M

Crocus Cloth, Midwest Abrasive Co., Owosso, Mich.

Emery Cloth, Evenrun Brand 3M, St. Paul., Minn.

Emery Cloth, Lightning Metalite Aluminum Oxide Behr-Manning Co., Troy, New York

Eraser, Gray, Blaisdell, BLUEZO No. 533T

Eraser, Red, Scripto, J160 Fineline

Eraser, Red, Scripto J190, Size No. 1

Eraser, White, Scripto, J160 Fineline

Hydro Micro Finish No. 5000 Plus Water

Stainless Wool, Brillo, Fine

3M TRI-M-ITE, Resinite Cloth, J wt., Closed Coat, Silicon Carbide

3M Scotch-Brite, Cleaning-Finishing Material, Very Fine Type A

ADHESIVES

(CHECK "SEALS AND SEALANTS" CATEGORY FOR DUAL FUNCTION PRODUCTS)

Adhesive A-4000, Armstrong Products Co., Bellflower, Calif.

Adhesive, Eastman 910

Adhesive Film HT424 American Cyanamid Co.

Adhesive - General Purpose, Rubber Base Loxite 703-487 MIL-A-5092B, Type III Firestone Tire & Rubber Co., Akron, Ohio

Adhesive, Heat Proof - Temp-Tite Pyrodyne Inc.

Adhesive Red RTV-106, General Electric Co.

Adhesive and Sealant, Kit #1c Epoxi-Patch Kit-Clear, Hysol Corp.

Adhesive and Sealant, Kit #6c Epoxi-Patch Kit Color, Aluminum, Hysol Corp.

Adhesive and Sealant, RTV-112 Self-Leveling, White, General Electric

Adhesive-System #5914 Los Angeles Standard Rubber, Los Angeles, Calif.

Adhesive, Translucent, RTV-108 General Electric

Catalyst "A" for Paper Type SR-4 Gage, Eastman

Catalyst "B" for Bakelite and Epoxy Type Gages, Eastman

Cement, Allen PBX, Mixed

Cement, Astroceram Precoat Type PC-1

Cement, Astroceram, Type B-LP

Cement, Bakelite, Baldwin, Lima Hamilton Corp.

Cement, BR 600, W.T. Bean Co.

Cement Duco, Dupont Co.

Cement, General Purpose MIL-C-4003 Amend. 1

Cement, Hitco No. 156, H.I. Thompson Co.

Cement Precoat, SR-4, Baldwin, Lima, Hamilton Corp.

Figure 3-3. Materials Compatible with Titanium Alloys (Sheet 1 of 17)

2-36. AIRPLANE CLEANING.

2-37. Introduction.

2-37A. The importance of keeping the airplane clean cannot be overemphasized. The use of approved cleaning materials and adherence to the recommended cleaning methods is essential. It is all too easy to overlook damaged parts when they are covered with dirt or grease. In addition to concealing evidence of damage, the accumulation of dirt and grease often contributes to failure. (Such an accumulation on moving parts can form a very effective abrasive which causes excessive wear in short order.)

2-37B. Salt water spray or film is especially injurious, and must be washed off as soon as possible after contact. It is important to realize that the airplane is subject to salt water contamination even though all over-water flight is at considerable altitude. Previous experience has shown that serious corrosion can result if the airplane is not thoroughly cleaned after such exposure.

2-37C. The following general recommendations are presented as a brief guide to better-ground-care of the airplane.

NOTE

Specific instructions on methods of cleaning the airplane as well as the approved cleaning solutions are provided in the next few paragraphs.

a. Consider airplane cleaning as a part of regular inspection and maintenance requirements. Clean airplane at regular intervals and whenever the need becomes apparent.

b. Use plugs, covers, etc., to protect the airplane at all times when it is on the ground.

c. Use only approved cleaning methods and materials.

d. Do not permit ground support or ground handling equipment to lean against or collide with the airplane.

e. Protect airplane surfaces with mats when work is being performed on or near them.

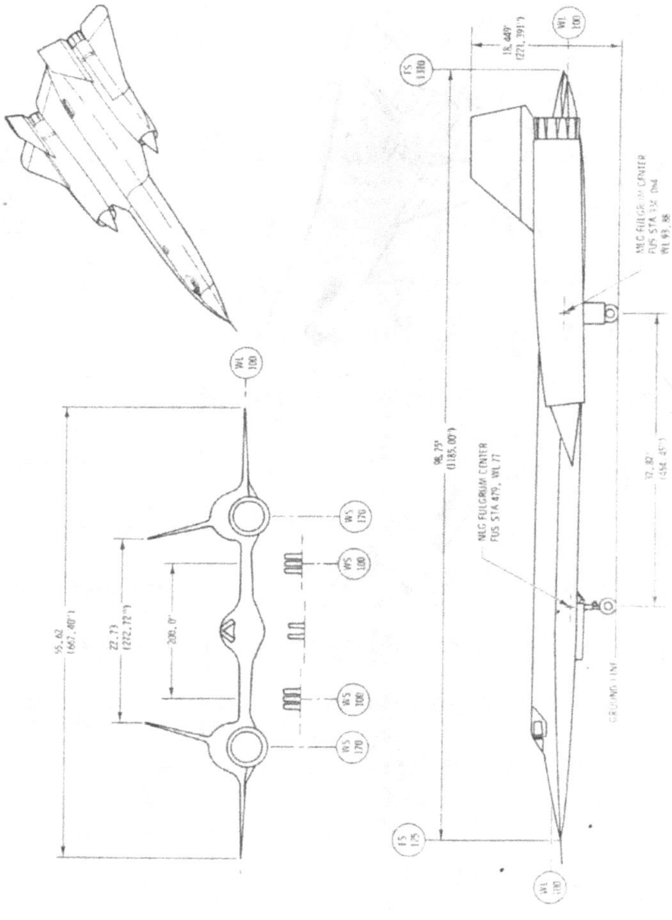

Figure 1-1. Airplane 3-View

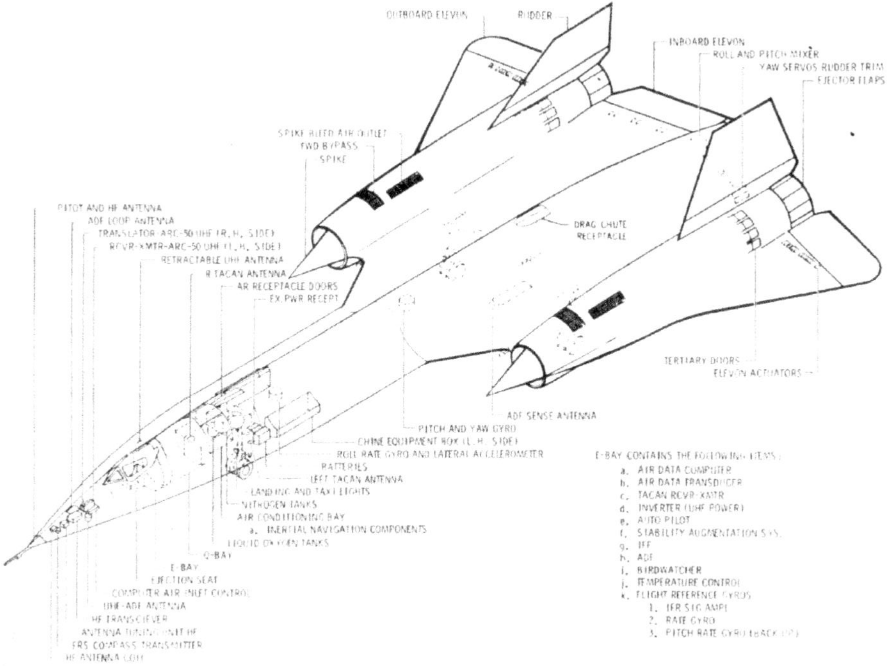
OUTBOARD ELEVON
RUDDER
INBOARD ELEVON
ROLL AND PITCH MIXER
YAW SERVOS RUDDER TRIM
EJECTOR FLAPS
SPIKE BLEED AIR OUTLET
FWD BYPASS
SPIKE
DRAG CHUTE RECEPTACLE
PITOT AND HF ANTENNA
ADF LOOP ANTENNA
TRANSLATOR-ARC-50 UHF (R. H. SIDE)
RCVR-XMTR-ARC-50 UHF (L. H. SIDE)
RETRACTABLE UHF ANTENNA
R TACAN ANTENNA
AR RECEPTACLE DOORS
EX PWR RECEPT
TERTIARY DOORS
ELEVON ACTUATORS
ADF SENSE ANTENNA
PITCH AND YAW GYRO
CHINE EQUIPMENT BOX (L. H. SIDE)
ROLL RATE GYRO AND LATERAL ACCELEROMETER
BATTERIES
LEFT TACAN ANTENNA
LANDING AND TAXI LIGHTS
NITROGEN TANKS
AIR CONDITIONING BAY
a. INERTIAL NAVIGATION COMPONENTS
LIQUID OXYGEN TANKS
Q-BAY
E-BAY
EJECTION SEAT
COMPUTER AIR INLET CONTROL
UHF-ADF ANTENNA
HF TRANSCIEVER
ANTENNA TUNING UNIT HF
FRS COMPASS TRANSMITTER
HF ANTENNA COIL
E-BAY CONTAINS THE FOLLOWING ITEMS:
a. AIR DATA COMPUTER
b. AIR DATA TRANSDUCER
c. TACAN RCVR-XMTR
d. INVERTER (UHF POWER)
e. AUTO PILOT
f. STABILITY AUGMENTATION SYS.
g. IFF
h. ADF
i. BIRDWATCHER
j. TEMPERATURE CONTROL
k. FLIGHT REFERENCE GYROS
1. IFR SIG AMPL
2. RATE GYRO
3. PITCH RATE GYRO (BACK UP)

2-44. Preparation For Cleaning.

2-45. Before cleaning airplane, proceed as follows:

a. If possible, park the airplane in a sheltered area or in the shade. If direct sunlight is encountered or high temperatures prevail, cool airplane external surfaces with plain water.

b. Attach low-resistance ground wire from the lug on the NLG gear to the grounding point in the service area. Grounding [illegible] placed in convenient locations on pad and aprons in open areas and in the pits in the hangars.

c. If complete airplane is to be cleaned, make sure canopy is down and locked; inflate the canopy seals or seal the canopy [illegible] to prevent cleaning fluids or water from entering the cockpit.

d. Close all doors, access panels, and exposed areas.

e. Install all necessary plugs and protective covers.

f. Mask or otherwise protect all electronic/electrical equipment which may be damaged or impaired by moisture.

g. Make sure all necessary cleaning materials and equipment are on hand.

2-46. Interior Surface Cleaning.

2-47. Airplane interiors, such as cockpit, equipment bays, and [illegible] sections which can be reached from doors or removable pa[illegible] be kept as clean as possible at all times. Corrosion can b[illegible] serious in these areas since they can be easily overlooked or neglected. Nuts, bolts, and small pieces of metal carelessly dropped a[illegible] neglected, can combine with moisture and dissimilar metal [illegible] up electrolytic corrosion. To keep these areas as clean as possible, the following precautions shall be taken each day:

a. Clean up spilled liquids immediately.

b. Inspect and make sure all drains and drain holes are open.

c. Protect airplane skins and structure from dents, scratches, and general damage.

2-47. (Cont'd)

d. Keep close watch on all tanks, reservoirs, and lines carrying liquids. Correct leaks as soon as possible. Do not permit leaking liquids to mix with water and pool up to create contaminants, fire hazards, or form dissimilar metal contacts.

e. Use a vacuum cleaner in the airplane, where practical, to pick up dust, dirt, and debris, especially in the cockpit area. Use a 1-inch wide brush like a feather duster to clean instrument panels, consoles, cockpit circuit breakers, instruments, etc. Use a cotton flannel cloth dampened in commercial Windex to clean instrument glass surfaces.

2-48. <u>Cleaning Exterior Surfaces of Aircraft.</u>

2-49. The exterior metal surfaces may be cleaned with a solution of warm soapy water, or with a mild solvent. The "warm soapy water" method is recommended for general over-all cleaning. The "solvent" method is reserved for removal of oils, greases, and residues in areas of heavy contamination where other cleaning methods are not suitable.

a. Mix the "warm soapy water" cleaning solution in a polyethylene plastic bucket. (About 2 gallon capacity.) Use 3 to 6 ounces of TURCO LEAK DETECTOR No. 5298 for each gallon of water used.

b. Use only <u>mild</u> approved solvents for the "hard-to-cleanse" areas.

c. Use the same <u>rinse</u> solution for both the warm water and solvent cleaning methods. Prepare the rinse solution by adding 1-1/2 ounces of C-153 solution to each gallon of water used. (C-153 solution can be obtained from airframe manufacturer.)

2-50. Warm-Soapy Water Cleaning Method.

NOTE

A work group of 3 teams of 3 men each is recommended. If possible start the entire group on same side of aircraft, one team at each end and third team in the middle, all working toward each other.

a. Mix cleaning solution as per paragraph 2-49.

2-47. (Cont'd)

d. Keep close watch on all tanks, reservoirs, and lines carrying liquids. Correct leaks as soon as possible. Do not permit leaking liquids to mix with water and pool up to create contaminants, fire hazards, or form dissimilar metals contacts.

e. Use a vacuum cleaner in the airplane, where practical, to pick up dust, dirt, and debris, especially in the cockpit area. Use a 1-inch wide brush like a feather duster to clean instrument panels, consoles, cockpit circuit breakers, instruments, etc. Use a cotton flannel cloth dampened in commercial Windex to clean instrument glass surfaces.

2-48. Cleaning Exterior Surfaces of Aircraft.

2-49. The exterior metal surfaces may be cleaned with a solution of warm soapy water, or with a mild solvent. The "warm soapy water" method is recommended for general over-all cleaning. The "solvent" method is reserved for removal of oils, greases, and residues in areas of heavy contamination where other cleaning methods are not suitable.

a. Mix the "warm soapy water" cleaning solution in a polyethylene plastic bucket. (About 2 gallon capacity.) Use 3 to 6 ounces of TURCO LEAK DETECTOR No. 5298 for each gallon of water used.

b. Use only mild approved solvents for the "hard-to-cleanse" areas.

c. Use the same rinse solution for both the warm water and solvent cleaning methods. Prepare the rinse solution by adding 1-1/2 ounces of C-158 solution to each gallon of water used. (C-158 solution can be obtained from airframe manufacturer.)

2-50. Warm-Soapy Water Cleaning Method.

NOTE

A work group of 3 teams of 3 men each is recommended. If possible start the entire group on same side of aircraft, one team at each end and third team in the middle, all working toward each other.

a. Mix cleaning solution as per paragraph 2-49.

Cement, Pyrodyne

Cement, Rubber, Carter's Clean Grip

Cement, Rubber, Sanford, No. 494

Cement, SR-4, Baldwin, Lima, Hamilton Corp.

Cement, Strain Gage, Brimor High Temperature Type U529

Ceramic Cement, Transonics 64CP

Epoxy Bond Adhesive Putty Atlas Mineral Products, Long Beach, Calif.

EPY-150, Baldwin, Lima, Hamilton Corp.

EPY-400, Baldwin, Lima, Hamilton Corp.

Glue-Transparent, T90CC1 MIL-A-4833 Sherman Williams Paint Co., Los Angeles, Calif.

Glue-White, Wilhold With Orthonol, Wilhold Glues, Inc., L.A.

Glyptal

Helix 385

Hysol - 151

Inspection Label, XS 1807

Pliobond - Goodyear

Red Cerro

Silicone Mix, Standard - 90% DC2106, 10% DC805, .16% Y-15 Catalyst

Silcastic 140, Dow Corning

Stabond C511, American Latex Co.

Tatnall GA-50

Tatro Cold Glue, A.H. Tatro Inc., L.A.

Template Adhesive, Pyrodyne

Thermofit Adhesive S1006

CLEANING AGENTS

Ammonia-Household, Lance Brand, Cloudy Dolly Adams Food Corp.

Basic H Organic Cleaner

Cal-Pac Chemical Soap #1190

Cleaning Agent - Panther 721, Panther Chemical Co., Fort Worth, Texas

Cleaner & Anti-Fogging Compound, Octagon Process Inc., Edgewater, N.J.

Cleaner, Bon Ami

Cleaner, Easy-Off

Cleaner, Kelite White Spray Mixed with Water (1:1)

Cleaner, Lectroetch Formula 2

Cleaner - Emulsion Y-415-E Yosemite Chemical Co. Los Angeles, Calif.

Cleaner - Liquid Solvent, Fine Organics Inc., FO-128 Rocwald Co., Brisbane, Calif.

Cleaner, Metal, Altrex 1097

Cleaner, Mild Acidic, LAC 32-266 Tec 901

Cleaner, Momode APC

Cleaner, Liquid - Amway Liquid Organic Cleaner, Amway Corp., Ada, Mich.

Cleaner, Oakite Vis Strip

Cleaner, Oakite Whistleclean D-12 Dissolved in Water

Cleaner-Optical Windows, Orvus W-A Paste, Procter and Gamble Co.

Cleaner, Phosphoric Acid Type, Kelite L-17A, LAC 32-260

Cleaner - Multipurpose, Amway Industroclean Amway Corp. Ada, Mich.

Cleaner, Prepaint TEC No. 901-7

Cleaner-Skin Waterless SBS-30, Sugar Beets Products Co., Saginaw, Mich.

Cleaner - Soap Calla 301, Midwest Supply Co., Napa, Calif.

Cleaner, Soap, Cal Pac Chemical No. 181

Cleaner-Steamite H, Cal-Tek Industries Los Angeles, Calif.

Cleaner - Swipe, Homcare Products, Houston, Texas

Cleaner-Tec #625, Tech Chemical Co., Monterey Park, Calif.

Figure 3-3. Materials Compatible with Titanium Alloys (Sheet 2 of 17)

Cleaner-Tergitol, Nonionic NPX, Union Carbide Corp.

Cleaner-Tergitol, Nonionic TMN, Union Carbide Corp.

Cleaner - Turco 3752 All Purpose Detergent Turco Products, Inc. Los Angeles, Calif.

Cleaner, Vitroclene 3881-3 Plus Additive 4215

Cleaner, X56 Welder's Spatter Off Chem X Brand

Cleanser - Bon-Ami Heavy Duty Institutional New Super Bleach Bon-Ami Co., N.Y. 22, N.Y.

Kein Mirror Cleaner, Precision Mirrors Corp.

Liqui-Nox Detergent, Alconox Inc., New York City, N.Y.

Machine Cleaner RV-41CL (White) RV-41CL Revised (Yellow)

Soap, Fels Naptha

Soap, Hand, Boraxo

Soap, Hot Solution Turco HTC

Soap, Liquid, Haas No. 2383, All Purpose

Windex with Ammonia D (New Hi-Speed)

DYES

Dye-Black No. DL8, Omega Paint Co. Los Angeles, Calif.

Dye, Oil Red O, Solvent Red 27 - Dissolved in Fuel, N.Y. Color & Chem. Co., Belleville, N.J.

Dye-White No. DL10, Omega Paint Co. Los Angeles, Calif.

Layout Dye, Steel Blue, Octagon Process Inc. edgewater, N. J.

Micro Supreme Green Dye

Micro Supreme White Dye

Micro Supreme Yellow Dye

Phenolphthalein, Lot #10442A Conray Products, New York, New York

Powder - Dyed Red Color, RN/491 Batch 51865-1, Semco Sales & Service Inc. Los Angeles, Calif.

ELECTRICAL

Kapton Wire Insulation, Dupont Plastic Haveg Wire

Sleeving - S8481 - 244, Insulectro Co., L.A. Calif.

Wire - Ray-Chem Spec. 44, MIL-W-81044/1 Ray-Chem Corp. Redwood City, Calif.

FIRE EXTINGUISHING AGENTS

Carbon Dioxide

Chemical Foam - Model FC 21

Municipal Water

FUEL

FWA 523 fuel

LUBRICANTS

Beeswax

Coating, Hi-Shear, Hi-Shear Corp.

Coating, Fluoride Phosphate

Compound, Anti-Seize, Go-Jo, No-Lok 72

Compound, Anti-Seize, Loctite

Compound, Drawing, Kern's DP1212

Crisco, Procter and Gamble Co.

Drawing Compound, Cycleweld DC2418, Chrysler Corp., Chemical Division

Dag 250 Lubricant, Acheson Colloids Co.

Drawing Compound, Cycleweld DC2431, Chrysler Corp., Chemical Division

Drawing Compound, Cycleweld DC2433, Chrysler Corp., Chemical Division

Figure 3-3. Materials Compatible with Titanium Alloys (Sheet 3 of 17)

Drawing Compound DP577 Kerns Pacific

Drawing Compound - Withrowdraw 535 (Was ACW 735) Arthur C. Withrow Co., L.A.

Dry Film Lubricant, Stretch Forming DP 3333, Kern's

Dus Lube

Easeoff 990, Texacone

Electrofilm 2006, Electrofilm Corp.

Electrofilm 4253, Electrofilm Corp.

Electrofilm 4396, Electrofilm Corp.

Electrofilm 4856, Electrofilm Corp.

Electrofilm No. 22-T, Electrofilm Corp.

Esnalube 382, Elastic Stop Nut Corp. of America

Everlube 1120-8 lubricant

Fafnir Lub FS171

Fel-Pro C-100, Fel-Pro Corp.

Fel-Pro Lub C-5, Fel-Pro Corp.

Fel-Pro Lub C-5A, Fel-Pro Corp.

Graphite Colloidal, in Lacquer, Dag Dispersion No. 41, Acheson Colloids Co.

Grease-Allube 602-M, Allen Corp.

Grease-Anderol L-758 Synthetic High Temperature, Lehigh Chemical Co., Chesterton, Md.

Grease, Darinax - Shell Multi-Purpose

Grease-Dow Corning DC4, Dow Corning Co.

Grease - Dow Corning 55M Dow Corning Co. Midland, Mich.

Grease, Dow Corning High Vacuum, Braun Chemical, Cat. No. 59293

Grease, ETRB, Shell Oil Co.

Grease, Gulf AF419 (no Additives)

Grease - AF 902, Gulf Oil Corp. of Calif.

Grease - AF903, Gulf Oil Corp. of Calif.

Grease, Mobil XRR15

Grease - Swepco Moly Grease #101, Heavy, Southwestern Petroleum Corporation, Ft. Worth, Texas

Grease, XRRB Mobil Experimental

Grease, 712167, Water Proof

Grinding Oil-Vantrol 915-M, Van Straaten Chem. Co., Chicago, Ill.

Grinding Oil-Withogrind 113 Arthur C. Withrow Co., Los Angeles

Hot Forming & Anti-Seize Comp. Everlube Corp., Formkote T-50, No. Hollywood, Calif.

Kaylube No. 3, Kaynar Mfg. Co. Inc.

Lano-Lube

Lubribond "A", Electrofilm Corp.

Lubri-Bond M, Electrofilm Inc., L.A. Calif.

Lubricant-Dow Corning 200 Fluid, Dow Corning Corp., Midland, Mich.

Lubricant-Armetco #1, Armetco Inc., Wooster, Ohio

Lubricant - Dri-Lube Air Industries Corp. Gardena, Calif.

Lubricant - Dri-Lube Thinner, Air Industries Corp., Gardena, Calif.

Lubricant - Everlube H 1120-8

Lubricant-Lusol (20:1 Water Solution, Anderson Chemical Co., Portland, Conn.

Lubricant-Moly Grease, Bemol Inc., Boston, Mass.

Lubricating Paste, Avdel (Chobert Rivets)

Lubriplate 130AA, Lubriplate Co.

LV 808 Grease, Laralube Co., L.A., Calif.

McKay Oil Plate, McKay Mfg. Co., L.A. Calif.

Miller's No. 120 G Lubricant

Molykote Loex 100, Alpha-Molykote Corp.

Molykote M-55, Alpha-Molykote Corp.

Molykote Type O, Alpha-Molykote Corp.

Pipe Thread Lubricant and seal, Slic-Seal Hole Inc.

Figure 3-3. Materials Compatible with Titanium Alloys (Sheet 4 of 17)

hfield Med. Hvy. RTO Oil

..ocol Molydisulphide Anti-Scuffing Oil, Straight

Sperm Oil

:okes Mechanical Pump Oil 3-391

Texaco Soluble Oil, Heavy Duty, SOHD PA 3319

PENETRATING OILS

DU'OL Penetrating Liquid Tool, American Parts Co., Inc.

Liquid Wrench, Radiator Speciality Company

Penetrating Oil, S'OK, Shamrock Specialities Co.
Texas City, Texas

Zyglo Penetrating Oil - ZL2, Magna-Flux Corp.

Zyglo Penetrating Oil - ZL22, Magna-Flux Corp.

PAINTS AND PRIMERS

Acrylic Plastic, Sinclair, Stuc-O-Lite, Sand Beige

rolite Cellulose Nitrate Lacquer, Type 2-Formula L-69,
ndrew Brown Co.

Catalyst Z6020, Mixed with Resin, Dow Corning Co.

Catalyst, Dow Corning Z6020 in 20% Iso-Propyl Alcohol

Catalyst, Pyrolac, Cat. No. TV-12

Enamel, Alkyd Semi-Gloss, Class A, Spec. TT-E-529A,
Color White No. 27875/Fed. Std. 595

Enamel, Black, Sperex VHT

Enamel, Blue, Sperex VHT

Enamel, Brown, Sperex VHT

Enamel, Gloss MIL-E-7729, Type I Andrew Brown or
Sherman Williams Paint Co's. L.A.
Red, Color No. 509/ANA Bull's 157 & 166
White, Color No. 17875/FED STD 595
Black, Color No. 17038 FED STD 595
Lt. Green, Color No. 14187/FED STD 595

Enamel - Gloss, Urethane Liquid Plastic, Violet,
Major Paint Co., Torrance, Calif.

Enamel, Gray, No. 16473, Spec. TT-E-489

Enamel, Gray, Sperex VHT

Enamel Green, Sperex VHT

Enamel, Heat Resistant Aluminum J-22977, 2362803

Enamel - Haddy High Temp. Gloss White, Aerosol, Stock
#533 Haddy Corporation, West Nyack, N.Y.

Enamel, Pittsburg, Sand Beige

Enamel - Red Gloss, Glyptal 1201 General Electric,
Schenectady, N.Y.

Enamel, Red, Sperex VHT

Enamel, Turquoise, Sperex VHT

Enamel, White, Sperex VHT

Enamel, Yellow, No. 13538, Spec. TT-E-489

Enamel, Yellow, Sperex VHT

Hi-Visibility Fluorescent System 8538 (173-Y-66)
Lemon Yellow, Fuller Paints

Hi-Visibility Fluorescent System 8533 (173-R-53) 633
Red-Orange Fuller Paints

Industrial Polyurethane Coating, CS2600 Flat Dark Gray
Lot A51-3B, Chem Seal Corp. of America, L.A., Calif.

Iron Blue - Color 2A, Hanline Bros. Baltimore Md.

Iron Blue - Color 2A, Sheffield Bronze Paint Corp.
Cleveland, Ohio

Lacquer, Black, Fullers MIL-L-6805C

Lacquer, Black, Krylon, No. 1601

Lacquer, Brown, Krylon, No. 2501

Lacquer, Gray, No. 16473, MIL-L-7178

Lacquer - #601-37875 Insignia White Per L-804 to Spec.
T-T-L-20 Per MIL-L-6805C, Andrew Brown Paint Co.
Los Angeles, Calif.

Lacquer, Orange, Krylon, No. 2401

Lacquer, Marking Fuller Identification TL 901 Pink

Lacquer, Martin Orange, No. 15-L-292

Lacquer, White, Krylon No. 1501

Figure 3-5. Materials Compatible with Titanium Alloys (Sheet 7 of 17)

acer, Pyrolac Cat. No. T-2-241

Rust Oleum, International Orange

Stain, Dykem, Black, DXX-553

Stain, Dykem, Brown, DXX-299

Stain, Dykem, Purple, DKG

Stain, Dykem, White, DW

Stain, Dykem, Yellow, DLT

Stencil, Ink, Reynolds Yellow (Spray Can)

Traffic Lac. Paint-Yellow, Sinclair Paint Co.

Varnish, Electrical Impregnating, General Electric SR-220

Varnish, Dow Corning 935, Xylene Solvent

White, Carter's Tempera Color

White, High Heat, PT-17328

White Coating, Hari Vulc #772-2 MIL-C-27227

White, Coating, PT401 Resin, No. 27875 (with H-11 Catalyst), Products Technique Inc.

PACKING MATERIALS

Bag - Nox-Rust Clear Pak 200 7/60 22019 (AER) Daubert Chemical Co. Chicago, Ill.

Bags, Plastic Lined, Guardian Paper Co., Milgard 1 Mil-B-121A, Grade A, Type I, Class I

Bag-Polyethlene, Wm. M. Frederics Co., L.A., Calif.

Barrier Paper - Oily Black MIL-B-121B

Cardboard Box PPP-B-636C

Cardboard Sleeving PPP-T-495a (1)

Cellu-Cushion N-200 Wadding Spec. PPP-C-843A Type II National Fiber & Cushioning Corp., San Dimas, Calif.

Charcoal Polyester CPR 9811-2.5

Cushion-Pak Cellulose Wadding, P3050, Line Embossed Kimberly-Clark, Laverne, Calif.

Foam-Polyether, Alpco #30

Heat Shrinkable Films Scotchpak Polyesther Film 45A48 MIL-P-22191 (AER) Type II and III Minnesota Mining & Manufacturing Co. St. Paul, Minn.

Kempak PPP-C-8436

Neutral Kraft Paper

Packing Material Polyether, UU34 Nopco Chemical Co., Newark, New Jersey

Paper, Dry Waxed, Acme Paper Co., San Francisco, Calif.

Plastic Sheeting - Transparent L-P-378a (2)

Polyesther Foam-Charcoal TF-Zero Manufacturing Co., Burbank, Calif.

Polyethylene Film Visqueen, Ethyl Corp., Baton Rouge, La.

Poly Induwrap, creped MIL-B-121B, Grade A, Type II, Class 5, Ludlow Papers

Shipping Bag No. 0 PPP-S-30 (1) Jiffy Mfg. Co. Hillside, N.J.

Strapping-Polypropylene, Avistrap FMC Corp. American Viscose Div. Industrial Packaging Dept., Philadelphia, Pa.

Scrim Foil-Continental Can Col, Inc. MIL-B-131D, Class 1, 2 Silver Gray, Noland Parker Co., Inc. Buena Park, Calif.

Scrim Foil-Ram Industries, Inc. FR2160 MIL-B-131D, Class 1, 2 Silver Gray

Scrim Foil Marvellum Co., MIL-B-131D, Class 1, 2 Silver Gray, Marvelseal 1311-D

Wrapping - Non-Rust Vapor Wrapper (Volatile Corrosion Inhibiting Treater) Instant Acting, Daubert Chemical Co., Chicago, Ill.

Wrapping - Non-Rust Vapor Wrapper MIL-P-34208 Type I, Class I, Style A, Daubert Chemical Co., Chicago, Ill.

Wrapping - Non-Rust Vapor Wrapper MIL-P-34208 Type I, Class 2, Style C, Daubert Chemical Co., Chicago, Ill.

Wrapping - Nox-Tarnish (Prevents Sulfide Attack On Metal) Daubert Chemical Co., Chicago, Ill.

Wrapping Paper - UU-P-286C (1)

Figure 3-3. Materials Compatible with Titanium Alloys (Sheet 9 of 17)

PLASTICS

Fiberglass Sleeving Treated With DC805 Plus Ti O2 White Pigment Bentley-Harris Mfg. Co., Conshohocken, Pa.

F.R.P. Silicone Glass Fabric MIL-P-997, Type GSG

Heat Shrinkable Sleeving, Polyolefin Ray Clad Thermo-fit-RNF100, Type 2, Non-flame Retarded Clear

Modified Teflon FEP Tubing

P080 Preimpregnated Cloth, U.S. Polymeric Co.

Parmacel ER 6016 Polyurethane

Plastic Blanket Material, XC-A198, Hysol Corp.

Plastic Glove, All Purpose Hand Guard Code No. IR, Polyethylene, Carl Biggs Co., Santa Monica, Calif.

Plastic Molding Compound (Phenolic) MIL-P-10420, Class II

Plastic Pipe-Kralastic Lasco ABS

Plastic Pipe-Polypropylene Carlon

Plastic Sheet - Eastran Butyrate Transparent Products Corp., Los Angeles, Calif.

Plastic Sheet-Laminated Glass Cloth Polytetrafluoroethylene Resin MIL-P-19161

Plastic Tint-Gray CS 3369-7

Plastic Sheet-Laminated Thermo-setting, Cotton fabric base, Phenolic Resin MIL-P-15035, Type FBM - Mechanical Grade

Polyethylene Tubing, Insulation Supply Co. L.A.

Polyurethane Coating Hardener For X-500 Magna Coatings and Chemical Corp., L.A.

Polyurethane Coating X-500, Semi Gloss White 4-W-16, Magna Coatings & Chemical Corp. L.A.

Polyvinyl Alcohol Sheet

Thalco Polyester Laminate (red)

TFE-Flurocarbon Resin Single Coated Glass Fabric Quality 405A-116SC, Dupont

Tubing - Polyethylene, Zipper ZT0500P or ZTZ0500P Zipper Tubing Co. L.A. Calif.

Tubing #66-P 3/8 Poly-Flo, Polyethylene Black and Clear, Ducommun

Tubing-Imperial Poly-Flo, Polyethylene Ducommun

Tubing-PTFE 4201, Irvington, Milky White

3M, Plastic Coating EC968

Uralane 5711 (Clear)

RESINS

Den 438 Plus MNA, Dow Chemical Co.

Den 438 Plus MNA Plus 10%, Antimony Oxide, Dow Chemical Co.

Epolite 90C Resin Plus 932C Hardener

Resin-Polyester, #S6-LS, Fibre Resin Corp. Burbank, Calif.

Sauereisen No. 29 Filler and Binder

Sauereisen No. 31 Filler and Binder

SC1013 Prepreg., Monsanto Chemical Co.

75% Laminac 4233, 25% Laminac 4146, 5% Antimony Oxide, 2% BZP Catalyst

RUBBER AND RUBBER GOODS

Amber Latex Tubing, Braun Catalog #56437 Van Waters and Rogers, Inc.

Braid Fluid Hose, Flexprene, Mercer Rubber Co., Inc., L.A.

BUNA-N Rubber 8696A L.A. Rubber Co.

BUNA-N Rubber 8696B L.A. Rubber Co.

BUNA-N Rubber Sheet Atlas 5009 Firm

BUNA-N Rubber Tube-Black, Rubbercraft Corp. of Calif., Torrance, Calif.

BUNA-N Rubber Sponge, LAC 23-931 (MIL-C-3133) Grade SB41, Rubatex Co.

BUNA-N with Polyacrylate X666K, L.A. Standard Rubber Co.

Fluoronated Silicone Rubber, LAC-23-952

Hose-Abrasoflex Multi-Purpose Mercer Rubber Inc., Los Angeles, Calif.

Figure 3-3. Materials Compatible with Titanium Alloys (Sheet 10 of 17)

sulating Gasket-Syn. Rubber Asbestos Fiber, Durabla z. Co., Dist. by Warren & Bailey, L.A.

MAT - Mitchell Air Strip, Natural Rubber, Valley Paper & Chemical Co., 2706 W. Burbank Blvd., Burbank, Calif.

Rubber Mount-J3058, Lord Mfg. Co., Erie, Pa.

Rubber Seal-B.F.B.-Watt Rubber Lot 5812A Goodyear Tire & Rubber Co., Akron, Ohio

Rubber Sheet - Buna N MIL-R-6855-1 40 Shore Rubbercraft, L.A.

Rubber Sheet - Silicone, Self-Adhering Cohrlastic 400, 500, 600, & 700, Connecticut Hard Rubber Co., New Haven, Conn.

Rubber Sponge - Silicone, Self-Adhering Cohrlastic R-10-70 Connecticut Hard Rubber Co., New Haven, Connecticut

Rubber Tubing-Silicone, Ben-Har 1258-2, Black, Blue, Green, Orange, Red, Violet and Yellow Bentley, Harris Mfg. Co., Conshohocken, Pa.

Rubber Tubing - Silicone, Durasyl, Ben-Har 1258-1, Bentley Harris Mfg. Co., Conshohocken, Pa.

Rubber Tubing-Silicone, Ben-Har 1258-X Transparent, ed Oxide, Bentley Harris Mfg. Co., Conshohocken, Pa.

Silicone Sponge-Molded #5000 Red, Soft. Los Angeles Standard Rubber, Inc., L.A.

Shock Cord-Unstretched Elastic Exerciser Cord, Spec MIL-C-5650

Silastic 860 RTV Silicone Rubber

Silicone Rubber, Closed Cell, LAC924

Sleeving-Silicone Rubber Coated Fiberglass Hygrade SR398-1 Markel

Stretch Tape-Pt No. SA 01020, Glass Supported Silicone Rubber Tape, Stauffer Chemical Co.

SEALS AND SEALANTS

(CHECK "ADHESIVES" CATEGORY FOR DUAL FUNCTION PRODUCTS)

Adhesive and Sealant, GE RTV-112, White, Self-Leveling, General Electric Co.

Buna N Rubber LAC 1-781E

Edge Sealer, Scotchcal, Number 3960, 3M Co.

Insulating and Sealing Compound, GE SS4005, General Electric Co.

Loctite Grade C, Loctite Corp., Newington, Conn.

Loctite Sealant Type H, Loctite Corp., Newington, Conn.

LTV-5006, General Electric Co.

Oyltite - Stik, Lake Chemical Co. Chicago, Ill.

Rubber, Mosites 651-60

Rubber, Silicone, RTV-102, General Electric Co.

Seal Coat, GB 103

Sealant - Aeroseal Aero Research Instruments, Chicago, Ill.

Sealant Catalyst, Thermolite 12, Metal & Thermolite Co., L.A., Calif.

Sealant, Dow Corning 92-015

Sealant, Dow Corning 92-018

Sealant, Dow Corning Q94-002

Sealant, Dow Corning Q94-003

Sealant, EC 847, 3M Co.

Sealant, EC1948, 3M Co. (Now EC2277)

Sealant, Fireproof Insulation Stabond 15-4 LAC Spec. 40-745

Sealant RTV 60, General Electric Co.

Sealant RTV 615A (Including RTV 615B catalyst), General Electric Co., Waterford, New York

Sealant RTV 630A (Includes RTV 630B catalyst), General Electric Co., Waterford, New York

Sealing Compound, Grey, 1249148, 3M Co.

Sealing Compound, Zinc Chromate Presstite Engineering Co., No. 590 Interchemical Corporation, El Segundo, Calif.

Sealing Compound, Zinc Chromate Putty Presstite Engineering Co., Interchemical Corp., El Segundo, Calif.

Viton B, Dupont Co.

XA-5117 Fillet Sealant

Figure 3-3. Materials Compatible with Titanium Alloys (Sheet 11 of 17)

SOLVENTS

Acetone

Alcohol Butyl TT-B-846b

Alcohol, Iso Propyl

Benzene

Water, Distilled

Water, Municipal

Methanol Plus Acetone (60:40)

Methyl Celosolve Acetate

Methyl Ethyl Ketone

Methyl Iso-Butyl Ketone

NAPHTHA, Shell Super V M & P

Retarder, Fuller's 8519, MIL-T-6095

Solvent, Dry Cleaner Spec P-D-680, Type 2

Solvent - Lektrik Spray Mechanical Supply Co., No. Hollywood, Calif.

Solvent, PT-1002, Products Techniques Inc.

Solvent Shell 140 Type 2

Solvent-Silicone Polish Remover Dupont 3919 Prep-Sol, Wilmington, Delaware

Solvent-Oakite 202, Oakite Products Inc. L.A., Calif.

Solvent, SR-4 Cement Precoat, Baldwin, Lima, Hamilton Corp.

Solvent, SR-4 Cement, Baldwin, Lima, Hamilton Corp.

Solvent, Stoddard PS661B, Type 2

Solvent, Straight Oil, Oxidized Petroleum

Solvent, Walker Super Kwick Rust

Thinner, Lacquer Dytex, No. 202

Thinner, Lacquer, LAC 41-683

Thinner, M 3 Lacquer, EC926, 3M Company

Toluene TT-T-548A

Water SW 5

Xylene TT-X-916 Grade B

TAPES

Brady Perma-Code Markers Stock No. A7-1-33 Tape No. 8-184

Crystal Bay Masking Tape, 3M Co.

Electrical Tape No. 27, 3M Scotch Brand, St. Paul, Minn.

Gum Tape-Olive Drab, (Permacel 69) JAN P-127, Type I, Grade B

Identification Tape-P996, L-T 101 (1), Type 3 Black, Blue, Green, Light Green

Identification Tape-P997, L-T-101 (1), Type 3 White, Yellow, Orange, Red

Masking Tape-Scotch Brand No. 232, 3M Co., St. Paul, Minn.

Masking Tape-Scotch Brand No. 238, 3M Co., St. Paul, Minn.

Mystik Tape - 7000 Mystik Adhesive Products Co., 1260 So. Central, Glendale, Calif.

Mystic Tape No. 5863, Spec. PPP-T-60 Type III, Class I, Red, Black or Olive

Mystic Tape Type 7402 SBKGB1

Paper Tape - Olive Drab

Permacel Mylar Tape #253

Permacel Tape P69, PPP-T-60 Type I, Class I, Red

Permacel Tape No. PE100

Permacel Tape No. 21

Permacel Tape No. 212

Permacel Tape P-703

Polyken Tape PPP-T-60B, Class I, Red, The Kendall Co.

3M Crystal Bay Transparent Tape

3M Electrical Tape No. 60 TFE

3M Electrical Tape No. 69

3M Electrical Tape - Scotch Brand No. 5, St. Paul, Minn.

3M Masking Tape No. 202 Fed. Spec. UU-T-106b, Creped, Type I

3M Pressure Sensitive Tape No. 700

Figure 3-3. Materials Compatible with Titanium Alloys (Sheet 12 of 17)

Silicone Adhesive Tape No. 361

Tape - Fluorolin 101A Joclin Mfg. Co., Wallingford, Conn.

Tape - Pipe Thread Sealant, 3M Scotch Brand #48

Tape - Polyesther Mylar Film, Clear Permacel P253[illegible]

Tape, Temp-R-Tape, Teflon, Connecticut Hard Rubber Co., New Haven, Conn.

Tuck Tape 90 TOD (Olive Drab) PPP-T-60, Type 3 Class I, Technical Tape Corp., New Rochelle, N.Y.

Tuck Tape No. 90 CD, PPP-T-60, Type I, Class I, Red

TEXTILES AND FABRICS

Aircraft Insulation, Type PF-105-700, Hi-Temp Corp., No. Hollywood, California

Cloth-HT1005, Hi-Temp Corp., No. Hollywood Calif.

Cloth-HT1026, Hi-Temp Corp., No. Hollywood Calif.

Cloth-8 oz. Unbleached, White Cotton Duck, Type III, Spec CCC-C-419

Cloth-Prima Cotton Canvas (60% Cotton, 40% Nylon), Dark Green

Cord-Clothesline Unfinished Cotton Sash Cord, Type A, T-C-571

Cord-Cotton Sash, LAC Spec 1-850

Cotton - Wing Socks, Black Rubber Impregnated Soles, Wm. M. Fredericks Co., Culver City Calif.

CRP 6217, 6228, 6229, Chem. Rubber Products Inc., Beacon, New York

Duck Brown Cotton, Waterproof, Kerosene and Wax Impregnated, Hume's Sporting Goods, Burbank, Calif.

Fabric-Nylon, Synthetic Rubber Coated (Red) Spec. AMS3274, CRP-6273, Chem. Rubber Products Inc., Beacon, New York

Fabric-Nylon Synthetic Rubber Coated (Brick Red One Side-Black Other) Spec. AMS3274, CRP6163, Chem. Rubber Products Inc., Beacon, New York

Felt-MIL-F-5030, Type I

Gloves - Grab It #62W, Edmont Inc., Coshocton, Ohio

Insulation, FPAA, Gray

Insulation, Slab 6-[illegible]

Lacing - Silicone Ty[illegible] Cord, Type 4[illegible], Dry Core, [illegible]" Dia. Varglas Co., [illegible], Rome, N.Y.

Rags, Wiping, Cotton Class 2, Grade B, White, Sandler Bros., L.A. F & W Was[illegible], Inc. L.A.

Thread-Barbours Pure Flax Sinew #6, White

Thread-Green, four 16 cord, Intrinsic Cotton Thread, Glaze 16

Trim Cloth, Painted Aluminized HT1032, Hi-Temp Insulation, Inc., No. Hollywood, Calif.

Trim Cloth, Painted aluminized - HTTOO SS1698 H.I. Thompson, Gardena, Calif.

Vivatex 31 - Preshrunk, Hume's Sporting Goods, Burbank, Calif.

WRITING AND MARKING MATERIALS

Ball Point - Red, Scripto Refillable Memo Med TX050

Crayon, Black, Scripto

Crayon, Red, Scripto

Crayon, Yellow, Scripto

Crayon Pencil, Blaisdell, Black 373T

Crayon Pencil, Blaisdell, Blue 368T

Crayon Pencil, Blaisdell, Red 365T

Dye, Identification & Layout, Dykem, Red, DX 296

Dye, Identification & Layout, Dykem, Steel Blue, DX 100

Dye, Yellow Aniline (alcohol fast) plus Butyl Alcohol

Felt Tip Ink Pencil, Ballerina Zip-Mark-Red

Fluid, Layout, Sprayon No. 603 Blue

Fluid, Layout, Purple, Great American Color Co.

Ink, Esterbrook Marking

Ink, Flomaster Black, Stock No. T-116

Ink, Flomaster White (Opaque)

Ink, Marsh Black Stencil, N-1

Ink, Marsh T-1 Black

Figure 3-3. Materials Compatible with Titanium Alloys (Sheet 13 of 17)

Ink, Marsh T-2 Red

Ink - Stamp Pad, Black, Class 1 American Writing Ink Co. Inc., Boston, Mass.

Ink, Stamp Pad, Black, Dickinson Ink Corp., Alhambra, Calif.

Ink, Water Soluble, Matthews, Red No. 8495A

Ink, Water Soluble, Matthews Red, Metal Marking No. 10646

Ink, Stamping, Independent No. 73X, Black

Ink, White Marking, Independent No. 73X

Ink Pencil, Eberhard Faber 705 No-Blot

Ink Pencil, National "Fuse-Tex" -7500

Ink Reconditioner, Independent No. 73X, Batch 3607

Kaimei Colour-Black

Kaimei Colour-Red

Labink-Black Batch No. 36080

Lead Fineline Red F-15

Lead, Scripto Red No. 1540

Marker, Esterbrook Yellow

Marker, Felt Tip Ballerina Zip Mark Green

Marker, Felt Tip Ballerina Zip Mark Orange

Marker-Felt Tip, Type A Magic Marker Black, Transparent, Speedry Chemical Products, N.Y.

Marker-Felt Tip Magic Marker, Green, Speedry Chemical Products, N.Y.

Marker-Felt Tip, Marks-A-Lot Black, Blue, Brown, Green, Orange, Purple, Red & Yellow, Carter's Ink Co.

Marker-Felt Tip Deluxe, Sanford's Black No. 1000-C, Bellwood, Ill.

Marker, Longlife (Red)

Marker-White 881, Carter's Ink Co.

Pen, Pentel-Black

Pen, Pentel-Blue

Pen, Wearever-Black

Pen, Wearever, Refill-Black

Pen, Wearever-Green

Pen, Wearever-Red

Stove Polish, Black Silk

Ultra Violet Ink

MISCELLANEOUS CHEMICALS AND COMPOUNDS

Acetic Acid

Ammonium Hydroxide

Anti-Skid Compound, Lynn Co. No. 16 Los Gatos, Calif.

Belt Dressing, Aquamatic Molecular Products

Black Chrome Bath, Corillum Corporation, Arlington, Va.

Cadmium Plate Stripper-Jell

Cadmium Plate Test Kit, Solution #1 - Ammonium Nitrate

Cadmium Plate Test Kit, Solution #2 - Sodium Sulfide

Cartridge Fumes, Amoco MCU-4/A

Casting Compound Hysol CU13

Cartridge Fumes, CM

Cerro Cast.

Cerro Tru

Chalk Pulverized White, Plus Water

Chromidizing Solution

Clay, Modeline #220, Gray, Binney & Smith, N.Y.

Coating - Humiseal 1B-15, Columbia Technical Corp., New York, New York

Coffee, Black

Corrosion Preventative Compound, Braycote 137D, MIL-C-16173C, Grade II

Cresylic Acid (Pure Phenol)

Davenite P12 Mica 325 Mesh Plus Water, John Rice Co.

Decal-Type SHR, Meyercord Co., L.A.

Figure 3-3. Materials Compatible with Titanium Alloys (Sheet 14 of 17)

Ink, Marsh T-2 Red

Ink - Stamp Pad, Black, Class 1 American Writing Ink Co. Inc., Boston, Mass.

Ink, Stamp Pad, Black, Dickinson Ink Corp., Alhambra, Calif.

Ink, Water Soluble, Matthews, Red No. 8495A

Ink, Water Soluble, Matthews Red, Metal Marking No. 10046

Ink, Stamping, Independent No. 73X, Black

Ink, White Marking, Independent No. 73X

Ink Pencil, Eberhard Faber 705 No-Blot

Ink Pencil, National "Fuse-Tex" -7500

Ink Reconditioner, Independent No. 73X, Batch 3607

Kaimei Colour-Black

Kaimei Colour-Red

Labink-Black Batch No. 36080

Lead Fineline Red F-15

Lead, Scripto Red No. L540

Marker, Esterbrook Yellow

Marker, Felt Tip Ballerina Zip Mark Green

Marker, Felt Tip Ballerina Zip Mark Orange

Marker-Felt Tip, Type A Magic Marker Black, Transparent, Speedry Chemical Products, N.Y.

Marker-Felt Tip Magic Marker, Green, Speedry Chemical Products, N.Y.

Marker-Felt Tip, Marks-A-Lot Black, Blue, Brown, Green, Orange, Purple, Red & Yellow, Carter's Ink Co.

Marker-Felt Tip Deluxe, Sanford's Black No. 1000-C, Bellwood, Ill.

Marker, Longlife (Red)

Marker-White 881, Carter's Ink Co.

Pen, Pentel-Black

Pen, Pentel-Blue

Pen, Wearever-Black

Pen, Wearever, Refill-Black

Pen, Wearever-Green

Pen, Wearever-Red

Stove Polish, Black Silk

Ultra Violet Ink

MISCELLANEOUS CHEMICALS AND COMPOUNDS

Acetic Acid

Ammonium Hydroxide

Anti-Skid Compound, Lynn Co. No. 16 Los Gatos, Calif.

Belt Dressing, Aquamatic Molecular Products

Black Chrome Bath, Corillum Corporation, Arlington, Va.

Cadmium Plate Stripper-Jell

Cadmium Plate Test Kit, Solution #1 - Ammonium Nitrate

Cadmium Plate Test Kit, Solution #2 - Sodium Sulfide

Cartridge Fumes, Amoco MXU-4/A

Casting Compound Hysol CU13

Cartridge Fumes, CM

Cerro Cast.

Cerro Tru

Chalk Pulverized White, Plus Water

Chromidizing Solution

Clay, Modeline #220, Gray, Binney & Smith, N.Y.

Coating - Huriseal 1B-15, Columbia Technical Corp., New York, New York

Coffee, Black

Corrosion Preventative Compound, Braycote 137D, MIL-C-16173C, Grade II

Cresylic Acid (Pure Phenol)

Daverite P12 Mica 325 Mesh Plus Water, John Rice Co.

Decal-Type SHR, Meyercord Co., L.A.

Figure 3-3. Materials Compatible with Titanium Alloys (Sheet 14 of 17)

Ink, Marsh T-2 Red

Ink - Stamp Pad, Black, Class 1 American Writing Ink Co. Inc., Boston, Mass.

Ink, Stamp Pad, Black, Dickinson Ink Corp., Alhambra, Calif.

Ink, Water Soluble, Matthews, Red No. 8495A

Ink, Water Soluble, Matthews Red, Metal Marking No. 10046

Ink, Stamping, Independent No. 73X, Black

Ink, White Marking, Independent No. 73X

Ink Pencil, Eberhard Faber 705 No-Blot

Ink Pencil, National "Fuse-Tex" -7500

Ink Reconditioner, Independent No. 73X, Batch 3607

Kaimei Colour-Black

Kaimei Colour-Red

Labink-Black Batch No. 36080

Lead Fineline Red F-15

Lead, Scripto Red No. E540

Marker, Esterbrook Yellow

Marker, Felt Tip Ballerina Zip Mark Green

Marker, Felt Tip Ballerina Zip Mark Orange

Marker-Felt Tip, Type A Magic Marker Black, Transparent, Speedry Chemical Products, N.Y.

Marker-Felt Tip Magic Marker, Green, Speedry Chemical Products, N.Y.

Marker-Felt Tip, Marks-A-Lot Black, Blue, Brown, Green, Orange, Purple, Red & Yellow, Carter's Ink Co.

Marker-Felt Tip Deluxe, Sanford's Black No. 1000-C, Bellwood, Ill.

Marker, Longlife (Red)

Marker-White 881, Carter's Ink Co.

Pen, Pentel-Black

Pen, Pentel-Blue

Pen, Wearever-Black

Pen, Wearever, Refill-Black

Pen, Wearever-Green

Pen, Wearever-Red

Stove Polish, Black Silk

Ultra Violet Ink

MISCELLANEOUS CHEMICALS AND COMPOUNDS

Acetic Acid

Ammonium Hydroxide

Anti-Skid Compound, Lynn Co. No. 16 Los Gatos, Calif.

Belt Dressing, Aquamatic Molecular Products

Black Chrome Bath, Corillium Corporation, Arlington, Va.

Cadmium Plate Stripper-Jeli

Cadmium Plate Test Kit, Solution #1 - Ammonium Nitrate

Cadmium Plate Test Kit, Solution #2 - Sodium Sulfide

Cartridge Fumes, Amoco MXU-4/A

Casting Compound Hysol CU13

Cartridge Fumes, CM

Cerro Cast.

Cerro Tru

Chalk Pulverized White, Plus Water

Chromidizing Solution

Clay, Modeline #220, Gray, Binney & Smith, N.Y.

Coating - Hariseal 1B-15, Columbia Technical Corp., New York, New York

Coffee, Black

Corrosion Preventative Compound, Braycote 137D, MIL-C-16173C, Grade II

Cresylic Acid (Pure Phenol)

Davenite P12 Mica 325 Mesh Plus Water, John Rice Co.

Decal-Type SHR, Meyercord Co., L.A.

Figure 3-3. Materials Compatible with Titanium Alloys (Sheet 14 of 17)

Deicer ID-110 (54) SOHIO Ice Melting Chem. SOHIO Chem., Cleveland, Ohio

Deicer Fluid MIL-A-8243A, Union Carbide Co., Texas City, Texas

Delchem D-2 Strip, 15AY Thick

Dip Coating 72-9-22, Textile Rubber and Chemical Co.

Descaler, Turco 4368

Electromark Electrolyte, Formula 20

Electromark Neutra Cleaner, Formula No. 1-A

Ethylene Glycol

For Oil

Flexane 85 Type

Flux, Kester #135

Flux, Thinner - Kester #104

Fly Spray, Bif Insect Killer, Wilco Co.

Freon C318

Gelatin Cadmium Plate Stripper

Glycol Crystals - Carbowax, Polyethylene Glycol #4000, Union Carbide Corp., Los Angeles, Calif.

Gull Droppings

Hand Cleaner RF 70

Hand Cream, Gabel's

Hand Lotion, Fend

Hand Protector - Kerodex 51 Ayerst Laboratories Inc., New York, New York

Hand Protector - Kerodex 71 Ayerst Laboratories Inc., New York, New York

High Alumnia Castable

Hi-Temp Liquid-Monsanto OS-138, Monsanto Chemical Co., St. Louis, Mo.

Hydro-Hone Rust Inhibitor #123

Insert Material, Esna SP-1, Elastic Stop Nut Corp. of America

Iridite Solution

Iron-Constantin Wire, Pyco Type JX-20-PW Pyrometer Co., Penndell, PA.

Iron Oxide Plus Water

Jet/Tec, Type A, Perforated, American Gas & Chemicals Inc., New York City, New York

Johnson's Wax Stick #140

Kerr Perfection Impression, Compound Tracing Stick #1 (Dental Wax)

Kirksite Scrappings

Kodak Rapid Fixer Solution A plus Water (1:1)

Lead Scrappings

Leak Detector-Oil Red "O", Solvent Red #27

Leak Detector-Channon 0531, Fluorescent Concentrate

Leak Detector - Tec #915 MIL-L-25567, Type I, Tec Chemical Co. Monterey Park, Calif.

Leak Detector-Turco #5298, Turco Products, Inc., Los Angeles, Calif.

Leak Tec Formula #372-A, American Gas & Chem. Co.

Leak Test Compound, MIL-L-25567A, (ASG) AMC Type I, Octagon Process Inc. Edgewater, N.J.

Magnesium Fluoride, 2% in 20% Nitric Acid

Mercury Vapor from Broken Fluorescent Tubes

Metl-L-Check Developer, D-70

Metl-L-Check Emulsifier, Remover E-50

MET-L-CHECK Dye Penetrant, VP30

Mirro-Grind

Mirror Glaze (MGH10)

Mold Release-Hysol AC4-4368, Pratt Tooling Materials Co., Paramount, Calif.

Mold Release, Miller Stephenson, S-142

Mold Release, Miller Stephenson, S-X-144

Monode Marking Process, T-2 Neutralizer

Monode Marking Process, T-10 Electrolyte

Figure 3-3. Materials Compatible with Titanium Alloys (Sheet 15 of 17)

Monsanto Corp., OS124

Nitric Acid, 10%

Oxalic Acid

Paraffin

Parting Agent Rezolin 633A, Rezolin, Inc., Santa Monica, Calif.

Penetrant DP-40, Dubl-Chek, Belmont Chemicals, L.A.

Penetrant Cleaner, Spotcheck Type SKC-NF, Magnaflux Corp., Chicago, Ill.

Penetrant Cleaner, Spotcheck Type SKD-W Magnaflux Corp., Chicago, Ill.

Penetrant Developer D-100, Dubl-Chek, Belmont Chemicals, L.A.

Penetrant Developer MIL-I-25135C Use with ZL-1B, -2 & -22 Penetrants only, Magnaflux Corp., Chicago, Ill.

Penetrant Developer, Spotcheck Type SKD-NF Magnaflux Corp., Chicago, Ill.

Penetrant Remover DR-60, Dubl-Check, Belmont Chemicals, L.A.

Permatex No. 3D Aviation, Form-A-Gasket, Permatex Company Inc., Brooklyn, N.Y.

Plaster, Casting Plus Water

Plaster, Densite Plus Water

Polyvinyl Alcohol Solution

Potassium Dichromate

Potassium Metaborate

Potassium Oleate (Pure)

Potting Compound, Durock "A" #0306, Physical Sciences Corp.

Protective Coating, Rust Lick 606

Protective Coating - Rustgone XF-165W Turco Products, Inc., Los Angeles, Calif.

Protective Coating - Rust Veto M.P., E.F. Houghton & Co., Phila., Penn.

Protective Coating - W.D. -40 Rust Preventative Rocket Chem. Co. Inc., San Diego, Calif.

Protective Cream, Industro, Ryder's Limited, Hollywood

Protective Dip Seal No. Spec. (Brown) 8.012-000-200

Protective Dip Seal White - 6.012-020-200

Putty, Aluminum, Devcon F

Rain Repellent, Boeing, Type I

Rain Repellent, Boeing, Type 3

Rain Repellent, K-394

Rain Repellent, K394, Plus Dibutyl-Butyl Phosphate (2:1)

RCD - Kodak MX509-2 Fixer, Goodyear Aerospace Co., Litchfield Park, Ariz.

Release Agent, Devcon

Reliance Tallowaid, Charles F. L'Hommedieu & Sons Co.

Rust Protective, Die, Ferr-Cote No. FS-103 Ferris Supply Company

Silver Plate Scrappings

Skin Conditioner Formula No. 212 Indco-Manufacturer

Smoke Signal, Distress Hand, Mark-1, Mod.1

Soapstone 200 Mesh Plus Water

Sodium Bicarbonate Solution

Sodium Dichromate

Sodium Phosphate Mono-Basic

Sodium Silicate Solution S35 (Water Glass)

Spray Graph

Spray Lat, Protective Coating

Spotcheck Dye Penetrant, Type SKL-HF, Magnaflux Corp.

Stearall Machine Fluid, G117L (Turco)

Strippable Release Agent, FR5888, Fiber Resin Corp.

Stripper, Leeder Chem. Co., No. B44 MIL-R-25134

Stripper, Oakite

Figure 3-3. Materials Compatible with Titanium Alloys (Sheet 16 of 17)

Stripper, Calcite "M"

Stripper, No. 651 W, Leeder Chemical Co.

Surmen 73-30

Sweeping Compound, Klin Chemical Co., San Francisco; Spec. P-S-863A, Type I

Sweeping Compound, Oil Dry Corp. of America, Chicago, Ill.

Sweeping Compound, Spec P-S-863A Type I, Continental Chemical Co. Sacramento, Calif.

Tin Plate Scrappings

Tooling Stone Compound T-10, Mixed with Water

Tri-Potassium Phosphate

Tri-Potassium Phosphate Additive (1%) Plus Water

Trisodium Phosphate

(3 Parts) Tri Sodium Phosphate (1 Part) Formamide, Dissolved in Water

Turco Precoat, LAC 37-674

Warner Block Repair Heavy Duty Sealer

Zyglo Developer ZP4, Magnaflux Corporation

Zyglo Developer ZP5, Magnaflux Corp.

Zyglo Emulsifier ZE3, Magnaflux Corp.

Zyglo Emulsifier ZE-4, Magnaflux Corp.

X-L- Hand Cream, Valley Paper and Chemicals Co., Inc.

Figure 3-3. Materials Compatible with Titanium Alloys (Sheet 17 of 17)

A-12s/ SR-71s ON DISPLAY:

Alabama:

A-12 #06930 - On display at the U.S. Space & Rocket Center, Huntsville, AL.

A-12 #06937 - On display at the Southern Museum of Flight, Birmingham, AL.

A-12 #06938 - On display at the USS Alabama Battleship Memorial Park, Mobile, AL.

Arizona:

SR-71A #17951 - On display at the Pima Air Museum, Tucson, AZ.

California:

A-12 #06924 - On display at the Blackbird Airpark in Palmdale, CA.

A-12 #06927 - On display at the California Science Center in Los Angeles, CA.

A-12 #06933 - On display at the San Diego Aerospace Museum, in San Diego, CA.

SR-71A #17955 - On display at the AFFTC Museum, Edwards AFB, CA.

SR-71A #17960 - On display at the Castle Air Museum near Atwater, CA.

SR-71A #17963 - On display at Beale AFB, CA.

SR-71A #17973 - On display at the Blackbird Airpark, Palmdale, CA.

SR-71A #17975 - On display at the March Field Museum, March AFB, CA.

SR-71A #17980 - On display at NASA's Dryden Flight Research Center as #844.

Florida:

SR-71A #17959 - On display at the USAF Armament Museum, Eglin AFB, FL.

Georgia:

SR-71A #17958 - On display at the Museum of Aviation, Robins AFB, GA.

Kansas:

SR-71A #17961 - On display at the Kansas Cosmosphere and Space Center, Hutchinson, KS.

Louisiana:

SR-71A #17967 - On display at the 8th Air Force Museum, Barksdale AFB, LA.

Michigan:

SR-71B #17956 - On display at the Kalamazoo Aviation History Museum in Kalamazoo, MI.

:

\

SR-71A #17964 - On display at the Strategic Air and Space Museum, near Ashland, NE.

New York:

A-12 #06925 - On display at USS Intrepid Sea-Air-Space Museum in the New York City Harbor.

Ohio:

YF-12A #06935 - On display at the National Museum of the United States Air Force, Wright-Patterson AFB, OH.

SR-71A #17976 - On display at the National Museum of the United States Air Force, Wright-Patterson AFB, OH.

Oregon:

SR-71A #17971 - On display at the Evergreen Aviation Museum in McMinnville, OR.

Texas:

SR-71A #17979 - On display at the USAF History and Traditions Museum, Lackland AFB, TX.

Utah:

SR-71C #17981 - On display at the Hill Aerospace Museum, Hill AFB, UT.

Virginia:

SR-71A #17968 - On display at the Virginia Aviation Museum in Richmond, VA.

SR-71A #17972 - On display at the Smithsonian National Air and Space Museum Steven F. Udvar-Hazy Center in Chantilly, VA.

Washington:

M-21 #06940 - On display at the Museum of Flight, Seattle, WA.

***United Kingdom*:**

SR-71A #17962 - On display at the Imperial War Museum in Duxford, England.

Virginia

A-12 #06931 – On Display CIA, Langley, Virginia,